WASHOE COUNTY LIBRARY
3 1235 03460 1547
P9-CIT-106

THE DOG WHO COULDN'T STOP LOVING

ALSO BY JEFFREY MOUSSAIEFF MASSON

The Face on Your Plate: The Truth about Food

Altruistic Armadillos, Zenlike Zebras: A Menagerie of 100 Favorite Animals

Raising the Peaceable Kingdom: What Animals Can Teach Us about the Social Origins of Tolerance and Friendship

Slipping into Paradise: Why I Live in New Zealand

The Cat Who Came in from the Cold

The Pig Who Sang to the Moon: The Emotional World of Farm Animals

The Nine Emotional Lives of Cats: A Journey into the Feline Heart

The Emperor's Embrace: The Evolution of Fatherhood

Dogs Have the Strangest Friends and Other True Stories of Animal Feelings

Dogs Never Lie About Love: The Emotional World of Dogs

Lost Prince: The Unsolved Mystery of Kaspar Hauser

When Elephants Weep: The Emotional Lives of Animals (with Susan McCarthy)

Kaspar Hauser (edited by Jeffrey M. Masson and Johannes Mayer)

My Father's Guru: A Journey Through Spirituality and Disillusion

Final Analysis: The Making and Unmaking of a Psychoanalyst

Against Therapy: Emotional Tyranny and the Myth of Psychological Healing

The Assault on Truth: Freud's Suppression of the Seduction Theory

The Complete Letters of Sigmund Freud to Wilhelm Fliess 1887–1904 (translated and edited by J. M. Masson)

A Dark Science: Women, Sexuality and Psychiatry in the Nineteenth Century

Santarasa and Abhinvagupta's Philosophy of Aesthetics (with M. V. Patwardhan)

Avimaraka: Love's Enchanted World (with D. D. Kosambi)

Aesthetic Rapture: The Rasadhyaya of the Natyasastra

The Oceanic Feeling: The Origins of the Religious Sentiment in Ancient India

Love Poems from the Ancient Sanskrit (with W. S. Merwin)

The Dhvanyaloka of Anandavardhana with the Locana of Abhinavagupta (translated by Daniel H. H. Ingalls, Jeffrey Moussaieff Masson, and M. V. Patwardhan; edited and with an introduction by Daniel H. H. Ingalls)

THE DOG WHO COULDN'T STOP LOVING

HOW DOGS HAVE CAPTURED OUR HEARTS FOR THOUSANDS OF YEARS

JEFFREY MOUSSAIEFF MASSON

HARPER
An Imprint of HarperCollins*Publishers*
www.harpercollins.com

THE DOG WHO COULDN'T STOP LOVING. Copyright © 2010 by Jeffrey Moussaieff Masson. All rights reserved. Printed in the United States of America. No part of this book may be used or reproduced in any manner whatsoever without written permission except in the case of brief quotations embodied in critical articles and reviews. For information, address HarperCollins Publishers, 10 East 53rd Street, New York, NY 10022.

HarperCollins books may be purchased for educational, business, or sales promotional use. For information, please write: Special Markets Department, HarperCollins Publishers, 10 East 53rd Street, New York, NY 10022.

FIRST EDITION

Designed by Jaime Putorti

Illustrations courtesy of istockphoto.com

Library of Congress Cataloging-in-Publication Data

Masson, J. Moussaieff (Jeffrey Moussaieff), 1941–
The dog who couldn't stop loving : how dogs have captured our hearts for thousands of years / Jeffrey Moussaieff Masson. — 1st ed.
p. cm.
Summary: "A groundbreaking, inspiring, and deeply personal exploration of the unique relationship between dogs and humans, from the bestselling author of *Dogs Never Lie About Love*"—Provided by publisher.

ISBN 978-0-06-177109-5 (hardback)

1. Dogs—Behavior. 2. Emotions in animals. 3. Human-animal relationships. I. Title.
SF433.M328 2010
636.7—dc22 2010006716

10 11 12 13 14 OV/RRD 10 9 8 7 6 5 4 3 2 1

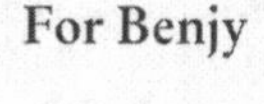

For Benjy

CONTENTS

PREFACE

For the past twenty years, I have been striving to see the world from an animal's point of view—sometimes I think I start to feel like one of them! Several books about the emotional lives of animals have resulted from this journey. Some have dealt with animal emotions in general (*When Elephants Weep*), others with the emotional life of a particular species (*Dogs Never Lie About Love, The Nine Emotional Lives of Cats*).

When I wrote about farm animals and the problems with domestication (*The Pig Who Sang to the Moon*), I was attempting to see if the insights I had gained about wild animals and companion animals could also be applied to domesticated animals we kept on farms. Did they too have emotional lives of some depth, I wondered? In *The Emperor's Embrace* (retitled in paperback *The Evolution of Fatherhood*), I tried to tackle the issue of how evolution played out in fatherhood among animals, but began to feel there was something eluding me. I turned to domesticated companion animals (dogs, cats, rabbits, chickens, and rats) for my book *Raising the Peaceable*

Kingdom, but again I felt something important was there just beyond my grasp.

Something lay at the heart of domestication that I could feel but could not yet put into words. I tried to look at the darker aspects of domestication in my most recent book, *The Face on Your Plate*, and I think I did succeed in understanding the unhappy lives of so-called food animals. In a scattered attempt to put my finger on the amazing diversity that characterizes the animal world, I wrote a book about my 100 favorite animals (little realizing how much work was involved), called *Altruistic Armadillos, Zenlike Zebras*. It was fun to write, and instructive, but still, something was missing. There had to be more; there had to be something that was different, that gave some evidence of the better impulses in the human species, something that explained our empathy, our sympathy, and even our compassion for members of a different species. I knew there had to be, because the feelings I had for some of the animals I lived with were unusually intense. When I thought back on these feelings, I saw that they almost always involved dogs.

Ever since I was a child, and continuing as an adult, I had puzzled over, and also marveled at, the relationship I (and others like me) had with dogs. Was there, in fact, something special about this, and only this, relationship? And if so, could it have a historical dimension? That is, could it be ancient, with roots reaching far back into human history? It seemed pure speculation, until a series of recent discoveries opened up the possibility that dogs and humans had a shared history of, at minimum, 15,000 years, and possibly far beyond. Many scientists had settled upon 40,000 years as the most likely compromise date for the beginning of relations between humans and wolves turning into dogs.

Whatever the date, it seemed something quite extraordinary happened to the human species, propelling us on a warp-speed evolutionary path that has taken us to the present. It struck me as remarkable, and highly suggestive, that at the very moment when humans were domesticating wolves into dogs, humans themselves were still in the throes of domestication, shedding their biological skins and being transformed by the culture they were building around them. This raised the striking possibility that it was partly through our association with dogs that we went from primitive humans to *Homo sapiens.*

The Harvard biologist E. O. Wilson has proposed what he calls the biophilia hypothesis: that we possess an inborn, species-typical tendency to affiliate with other animals. I agree that we do make these connections with other animals and derive great pleasure from so doing. We are unique in this respect. But are we wholly unique? As I thought about this question, I was finally able to put my finger on what had teased and eluded me for so long: that there was something different about one animal species, something they shared with us more than they did with their fellow nonhuman animals. That species was the dog, and what they shared with us so uniquely was their capacity for love. So yes, another book on the emotional lives of animals—a book that takes E. O. Wilson's hypothesis and gives it a new twist. Not only have dogs and humans influenced each other for tens of thousands of years, they have done so in far more profound ways than any other two species on earth, and primarily in the direction of acting on the capacity for love in all its different manifestations, such as showing sympathy, feeling empathy, and expressing compassion.

What does it mean to say "dogs make us human"? It is a phrase that has been used on several occasions (for example, by

Temple Grandin) in the last few years. I think it means something like: Through associating with dogs, we went from primitive humans to *Homo sapiens*. What would be a good analogy? That through association with humans, *Canis lupus* (wolf) became *Canis lupus familiaris* (dog). It worked both ways.

This is a hypothesis. It is the nature of this problem that it can never be proven. More evidence may make it more likely, but its value will need to be restricted to its evocative and explanatory power. Does it help us understand certain human and canine mysteries? Many people bond with dogs in more profound ways than with any other species. Why is this bond—and this bond only—unique among all the attachments humans are capable of forming, parallel only to the attachment we feel for our own children?

Why, of all animals, does the dog follow us, or rather accompany us, wherever we go on earth, from the tropical rainforests to the frozen icescape of the far Arctic? Why are there no tribes, no countries, no societies, not even villages, without dogs? Nowhere in the world are dogs absent.

Why is it that people have the capacity to befriend animals who are willing to take a chance and come into our intimate orbit, when no other animal on earth will do so—only dogs? Because a long time ago, in the words of Konrad Lorenz, man met dog. (Though it was probably more likely that long ago, 40,000 years ago or more, woman met dog. To this day, dogs by and large prefer women to men.)

The great Stanford anthropologist Richard Klein has long argued that something very significant happened to humans about 50,000 years ago. It has been called the "dawn of human culture," or "human revolution," a "creative explosion," or a "sociocultural big bang." He explains that the relationship between anatomical and behavioral change shifted abruptly about 50,000 years ago. Everything was suddenly put on fast-forward: language, culture, and the ability to innovate. Something, he thinks, changed at a neural level. What was new? Suddenly there are cave paintings where there had been none before, in France, in Spain, in Italy, but also in the Middle East and Australia. There was color—vivid, bright colors. There were rainbow serpents, for example, when before there were none. There were sculptures. Even human language, the way we know it today, sophisticated and capable of a million permutations, may have reached its apogee then. Weapons became more sophisticated. Beads that were hard to make were used as symbols. The very *idea* of a symbol belongs to this time. Suddenly humans buried their dead, and often did so elaborately. Put a suit on the man from 50,000 years ago and have him walk down Fifth Avenue. Nobody would stare. Let him speak and his ideas would not appear strange. He is us. Nor would he be surprised to see us accompanied by a dog. Walking with a dog would be entirely familiar to him. Is it possible that all of these innovations, these civilization-changing advances, had to do with our first encounters with wolves or wolves who were already slowly turning into dogs—that is, animals who share our lives?

It may sound unremarkable, but when you think about it, it is truly a phenomenal thing. No other animal behaves like a dog. I have always maintained that nothing in real life can compare to the vividness, the strangeness, the beauty, and the excitement of our dream life at night. Except dogs. Because everyone who lives with a dog is constantly in a dream to some extent. How is it possible that this creature, that looks so unlike us, is nevertheless asleep in our bed, up with us in the morning, walking with us in the afternoon? Watching us with love in his or her eyes? It is truly astonishing. And I believe it has this effect on us almost constantly. It forces us to think thoughts we would never otherwise think. Thoughts about sympathy, and empathy, and companionship, and friendship, but above all, thoughts about love. With dogs we have begun the long process—by no means assured, by no means nearly finished—of recognizing the basic sameness, the identity of *all* sentient beings. The Buddhists had it right. But it was not to come from Christ, Moses, or the Buddha, but rather from that little guy, walking so trustingly next to you, who would never abandon you for anything. From him and only from him did we learn that we could cross the species barrier and love other life-forms. It is no coincidence that of all the animals on this planet, only two consistently love other animals: humans and dogs.

How did it happen? What does it mean? This is the journey I set out upon. Read on.

INTRODUCTION

BENJY, THE DOG WHO COULDN'T STOP LOVING

Long before any other animal or plant was domesticated, we began our domestication of the wolf into the dog. This is well-known, and began sometime between 130,000 years ago, when humans first became *Homo sapiens*—"wise people"—and 15,000 years ago. What is less often considered is that this domestication may well have been a two-way street—a mutual and profound transformation for both species. Wolves were created into something new by humans, but as this unfolded, humans were changed in kind. This is, evidence suggests, the first and perhaps the only instance of what some are calling mutual domestication.

Could it be that dogs made us fully human? Without the wolf, would we have become a different species? I think it is very likely. We have been so intimately tied to dogs for so many thousands of years that we have come to resemble one another in crucial ways that are simply inapplicable to any other two species on earth, now or in the past. As a result, we are also the only two species who readily become friendly with other species. Dogs make friends across the species barrier, and so do we. Yes, it is true that some other animals do this as well from time to time. But dogs and humans do it as a default, reliably and consistently. There is hardly a human on earth who has not at some point in his or her life felt close to an animal from a different species—and not just a dog. And almost every dog has at some point felt friendly feelings not just for us or for other dogs, but for other animals as well. If you stop to think about it, this is no small achievement. It must be more than mere coincidence that of all the species on earth, only humans and dogs have attained it.

IN THE MARK Twain Papers at the University of California's Bancroft Library, there is a copy of Charles Darwin's *On the Origin of Species*, with Mark Twain's penciled annotations. He marked a passage about the emotions of dogs: "In the agony of death a dog has been known to caress his master, and every one has heard of the dog suffering under vivisection, who licked the hand of the operator; this man . . . unless he had a heart of stone, must have felt remorse to the last hour of his life." Mark Twain was particularly touched when he read the autobiography *The Life of Frances Power Cobbe*, about England's leading antivivisectionist, who died in 1904. Cobbe writes about a visit from her old friend, John Bright. Bright told her about a very poor woman, severely disabled, who was confined to her bed while her husband sought day labor. A neighbor sometimes brought her food. She was entirely helpless, but she had one comfort: a beautiful collie who lay by her side and licked her crippled hands and fingers.

> *One summer he came to the cottage, and the hapless cripple lay on her palette still, but the dog did not come out to*

> *him as usual, and his first question to the woman was: "Where is your collie?" The answer was that her husband had drowned the dog to save the expense of feeding it. Bright's voice broke when he came to the end of this story, and we said very little more to each other that dinner.*

These are only stories, but they are true stories, and there are thousands of them, recounted over and over by people who love dogs, precisely because of this ability to love without seeking reward. It strikes humans as almost uncanny. It's easy to speculate that dogs evolved to love us like this because their survival depended upon it. But what did that evolutionary process do to us?

This is a book about dogs that questions what makes our relationships with them—and theirs with us—unique, and asks what these relationships have meant to us. Much has already been said on the subject in countless books and articles that pour out every year. Yet only recently have academics engaged in serious study of dog and human coevolution. Moreover, our intimacy with dogs is so profound that in certain respects we take it for granted. You have to step back to consider the full implications of the fact that it is quite unprecedented in the history of any species on our planet, over the last several millions of years, for such intimacy to have arisen. Our relationship with dogs predates all other domesticated species by tens of thousands of years. Dogs and humans coevolved for a significant portion of the total lifespan of our respective species. In crucial ways, dogs helped make us human.

Is it possible that humans owe our ability to empathize, and perhaps even love selflessly, to our long association with dogs? This may sound sentimental, but it is such only if it is not true. If it is true, it provides a clue to one of the most important questions about human nature ever asked. Where does our ability, our desire, our need to love come from?

There are times in life when we feel we are exactly where we want to be at that moment, and they often have to do with love. The moment seems frozen: not one that will be followed by another or will ever end. For me, that moment comes at around 6:30 nearly every morning, when both boys come to visit my wife, Leila, and me in bed for a morning cuddle. Within seconds, a dog named Benjy follows. It's a lovefest. Wherever he is sleeping in the house, once he hears Manu and Ilan coming into our bedroom, Benjy comes along, too. He jumps up on the bed and begins a morning dance of recognition. He kisses all of us over and over and then rolls over onto his back and makes funny faces and strange guttural sounds of joy. He reaches his paws out to each one of us, so that four of us are holding his paws. (Sometimes one or two or even all three of our cats join us—but I must admit that they are a bit embarrassed by the sentimentality of it all, and usually leave in a huff before it gets too lovey-dovey.) Benjy smiles. He laughs. He is in ecstasy. So are we. It is contagious. There is no need to interpret what he feels, for we feel it too. It is unbridled love and happiness, sufficient for that moment. Whatever follows in the day for him or for us, those few moments every morning provide a reminder of what pure happiness can feel like, and how it seems like a moment in time that never really ends. We are so used to the cliché of dogs living in the moment, but in that morning moment we are, all five of us, living in it. It feels entirely natural, but it is also, when you come to think of it, something of a mystery. Benjy the dog, or Benjy the wolf, and four humans, all engaged in an identical bonding activity that plays itself out over and over in thousands of households, and probably has for thousands of years. What, exactly, is going on here? What is this strange relationship we have with dogs, yet not with any other animal?

I began to ask these questions when my family adopted the latest in a long line of four-footed companions: Benjy, a failed guide dog. Benjy obeyed only four commands: Leave it, down, sit, stay. He didn't love walks. He seemed a bit dim. His favorite activity was for somebody to give him a rubber Kong (a hollow tube made of hard rubber) filled with treats and let him work on it for the next half hour. But when he is done and obviously ready for a refill, if told to "Bring the Kong," Benjy looks perplexed and utterly lost. The average guide dog must learn some 70–80 commands, so it is no wonder that Benjy failed, even though his "handlers" dearly wanted him to succeed.

But there is one area where Benjy excels: He cannot stop loving. He loves all dogs, all humans, all cats, all rats, and all birds. He loves them all equally and intensely. He has yet to meet a species he is not fond of. He is not extraordinary: He is a lab, after all. He was socialized before we adopted him. He was never hit or yelled at as far as I know. He has never been in a fight with another dog, although my three cats sometimes slap him in the face just for the sheer pleasure of it, and he always looks completely mystified. He is a big (80-pound), strong dog with huge teeth and an awesome jaw, but I have never seen him lose his temper, get angry or even testy. At most, if pushed (as when I tell him he has to come up the hill—we live on the beach, and the path leading up to the street is steep and long), he will put on his sad face and begin a glacial walk up the hill. It is the same look he gets when I tell him there is no more food for the night: resignation and the dim hope that a mistake has been made.

Wherever I go during the day, Benjy goes too. When I do my errands in town, he comes along, perfectly happy to wait for me outside the bookstore or the post office or the bank. But over the last few months, he has taken to finding his way to the interiors of these destinations, because he knows he's clearly welcome there. People like to see him. This is in part because he's got love written all over him. But it's also partly because he has a peculiar habit: He acts as if he has met many people before, even when he hasn't; that they are close friends, even when they aren't; that he spends time with them often, even when he doesn't. I must confess that I encourage this. I often say, "Benjy, look who's here!" He responds by frantically turning his head in every direction searching for the person he knows. Whoever happens to fall into his line of sight at that moment becomes the designated long-lost friend, and he rushes over. He looks up at the person with adoration, his whole body quivering with the excitement of seeing him or her "again." The funny thing is that he has never seen the person before in his life, nor has the person ever seen him. Yet both parties welcome this subterfuge. They both know it is false. That is, false only to a point, because it is also true: Benjy is delighted to see whoever it is, even if for the first time. And the person is rarely displeased to see a large dog with such obvious friendly intentions. If the person is a child, well, then the greeting is even more effusive and the parents feel that the child has been privileged to experience dog love in its purest form. I make it up; Benjy makes it up; yet the feelings are genuine. It is as if in seeing *any* human being, Benjy is seeing our entire species. Every person stands in for his most intimate pack. Benjy has succeeded in making the entire human species into his personal pack. In this pack there are no alpha or omega dogs or people, only pure equality, the egalitarianism of pure affection.

I DON'T TAKE any credit for Benjy's easygoing nature. Unlike a child, where parents can take a certain pride if their kid is friendly and gentle with other kids, Benjy didn't inherit or learn to love from me. I just got lucky. But whatever the source, Benjy's got it—the love bug, the love gene, the love need. Of course, breed and temperament are part of it. Much as we might like to believe that every dog has the potential to be as loving as Benjy, it would appear not to be true. As media stories of dog attacks remind us, they are sometimes seen as predatory wolves in a dog's clothing. Some dogs are aggressive from a very early age. Benjy was the opposite from puppyhood, and has remained gentle beyond most other dogs I have known.

The other day he was in his favorite spot—at school with the second-graders surrounding him and jumping all over him—when the teacher worried that should one of the children inadvertently hurt him, Benjy might snap. "Oh no," I explained. "Never. Benjy would never snap at any living creature, of that you can be certain." I have wondered and wondered about the source of Benjy's extraordinary gentleness. It feels like a gift. Everybody benefits from it. If Benjy recognizes somebody, he rushes up to that person and showers him or her with affection. He licks and his tail goes wild and he has a look of pure happiness on his face. Smile is not the right word; he is laughing with delight. But this behavior does not make Benjy an exception, for he is the poster child for the rule that dogs have a special capacity to love. He may be exceptional in the sheer intensity of his feelings, but in this he is merely a showcase of how and why dogs became a human's best friend.

CHAPTER 1

GETTING BENJY

At the time when Benjy came onto our radar, we were not ready for another dog. We were already a large family: my lovely wife, Leila; our two boys, Ilan, now thirteen, and Manu, now eight; my ninety-one-year-old mother (God bless her); three domesticated rats; and three attention-seeking cats. We did not know how much longer we would be living in New Zealand and I did not want to worry about caring for another animal. True, we lived in the ideal spot for animals (or humans, for that matter), but that's not reason enough to adopt an animal, even if you love them as we do.

Eight years ago, we were visiting New Zealand from London, where Leila, a pediatrician, was studying at the London School of Tropical Medicine. I was doing research for a book about the emotional world of farm animals. I had heard about a pig that lived on a beach just fifteen minutes outside Auckland, so we went down to visit. As we wound our way down a steep path through a Pohutukawa forest, I got my first glimpse of the expanse of the ocean and the green volcanic islands just offshore. At the bottom of the path were ten homes, all just a few yards from the water. I was hooked. I wanted to live on that beach. Attracted by the lack of cars (safe for children and animals), the magic beauty of the historic site (the treaty of Waitangi, the defining document in New Zealand between the Maori and the settlers, was signed on this beach), and a childhood love of the sea, I could not get it out of my mind.

And so six months later, we moved from London and built an eco-home made of cedar and glass right there on the beach. When we arrived in New Zealand, our first son, Ilan, was just four. Leila gave birth to our second son, Manu (a Maori word for *bird*) here in 2001. I sat down to write my books about animals, the kids went to local schools, and Leila began a practice for children on the autistic spectrum using the biomedical approach (no drugs, only supplements; no dairy or gluten). It was an all-around good time. Hell, it was as close to perfect as one could realistically expect.

Then a woman called me, a woman who knows there are things I find hard to resist. Her name was Belinda Simpson, and she had read my book *Dogs Never Lie About Love,* in which I explore the emotional lives of dogs and where I first wondered if there were areas of feeling in which dogs are simply our superiors. She wanted to come down to our beach to tell me about her work training seizure-alert dogs.

Belinda had suffered from severe seizures for many years. Her golden retriever companion, Bradley, seemed to know up to an hour beforehand when her seizures would come and would take measures to see that Belinda would not harm herself. He would gently herd her to a nearby couch or armchair. Then, during the seizure, he would keep his body tightly close to hers to prevent her from falling. Belinda began to train other dogs who showed a spontaneous talent for alerting to seizures. I was fascinated. "I did have one huge failure, though," she told me. "And it broke my heart. A dog came to me after failing to qualify as a guide dog for the blind, even though he had been bred to become a guide dog and had undergone much training. The officials at the Royal New Zealand Foundation of the Blind told me he just didn't seem to want to work. At the same time, he was one of the most agreeable and likeable dogs they have ever known."

Later I was to meet many of the people from Guide Dogs who had been involved in some way with Benjy, and all of them, to a person, remembered him. Why? Because he was so mellow, they all agreed. A love-dog, made, not for service, but for loving. Belinda discovered that Benjy was no better as a seizure-alert dog than he had been as a guide dog for the blind (more about that later), but she also found that they were right; he was remarkably affectionate. "He's one of the most loving dogs I have ever met," she told me. "Sadly, he's with a family now who can't keep him—his fourth family in two years."

Manu and Ilan were listening. "Please, please, Dad. Can we just go see him?" I reluctantly agreed, sensing what lay in our future.

Why was I reluctant to have a dog move in with us, loving dogs as much as I do? For one, we are a family that moves around a great deal. I could not be sure we would be in New Zealand for the next year, and I was afraid we might move to somewhere it could be difficult to bring a dog along. Dogs, like children, tie you down. Suddenly there is another being whose needs and desires must be taken into account and often take precedence over your own. For years. And years. Then dogs die. Well, we all die; get over it. But dogs die long before their owners. Our dog would probably die just when the children are moving out on their own, so Leila and I would be left bereft twice. It just doesn't seem fair that they have such short life spans. True, they live their twelve-odd years with such intensity that it really is like living to eighty-four. Still, they leave such a hole in our lives. Getting a new dog after the death of an old one seems almost like a betrayal. Was I ready for so much responsibility? Did it matter? No, of course not! As many parents know, dogs seem to come into one's life without obeying any of the standard rules of acquisition. One day you are living dog-free, and the next you are not.

Also, I must confess that I hesitated taking on an adult dog. Puppies see us *in loco parentis*. As they grow, we become all they know and all they have ever known. Not so for a two-year-old dog. Like a boundary-testing adolescent, he will have well-worn habits and no doubt a certain degree of attitude. And here was a dog who had had four homes—the proverbial foster dog. What did that mean? Maybe he was difficult; maybe each of the families found him such a challenge that they gave up and asked the Foundation of the Blind to find him another home? Was he an inveterate cat-chaser? Intolerant of small children? Aggressive to other dogs? After all, these are not such uncommon behaviors. Like humans, dogs can be twisted into some shapes that are pretty strange. Moreover, even if none of these turned out to be of concern, what about his experiences in his previous homes? Could it be that he had been hit or in some other way traumatized? Could he overcome such a past, and how long would it take?

Then there were the more mundane questions: What could we offer him? Were we prepared to take him for long daily walks? He was a big dog, already tending toward obesity. Could we make sure he got as much exercise as he needed? And what about obedience training? He had not made it as a guide dog. Surely one of the reasons must have been (and indeed it was) his reluctance to obey commands, to be told what to do. Nobody in our family was a no-nonsense, take-charge kind of person. Would he recognize this, take advantage, and pursue his own desires rather than bend to ours?

These are legitimate questions for anyone adopting a dog. Unfortunately, the answers are usually irrelevant to the final decision. Our children had already made up their minds the minute they laid eyes on Benjy. I would have intervened if I had been sure this was the hound from hell. But that he certainly wasn't—that much I knew. What I could not know, especially given his foster status and his failed professional pursuits, was that Benjy would turn out to be the dog who couldn't stop loving.

WE DROVE OUT to see the family who were taking care of Benjy temporarily, where he had lived for a few months. We saw a large golden Lab playing with the two other dogs already in the household. For a dog who had been moved from home to home over his short life (he was almost two when we first met him), he seemed remarkably poised and well adjusted. It felt like visiting a family whose child you are thinking of adopting. I was disappointed immediately in two things: He did not exactly rush to greet me as a long-lost friend. Why should he, since we had just met for the first time? Nevertheless, this was my fantasy; that he would recognize his "real" family and not leave our side until we relented and took him home. (This is not just fantasy. Dogs who are up for adoption in shelters often somehow know that each person who plays with them is a potential family.) Also, I expected a much sleeker, more athletic-looking dog. Nor did I like his big jowls, from which I sensed saliva would soon be dripping (I was wrong about this). I felt, I admit, a tad disappointed. The boys, on the other hand, were not the least bit put off by anything. They took to Benjy immediately. He was more demonstrative with them than with the adults, a trait that has remained to this day with Benjy (he has never met a child he does not like; no, strike that—*love* is the more appropriate word here).

When I said I would like to take Benjy for a walk, did I imagine the two women who were his caretakers exchanging a slightly worried glance? We did not know it at the time, but Benjy was on Prozac and did not like to walk. *Prozac? Oh, no,* I thought. I hate psychiatric drugs for people, let alone dogs. I had heard that when Prozac was first tested on dogs, even before the drug was approved for such use, it sometimes made them vicious; others it made lethargic. But "all right" they said, and we set out on our walk. Within a few minutes, we came to a small corral by the side of the road. At the far end were some miniature horses, and one of them came galloping up to the fence. Benjy pulled away from his leash, and for one terrible second I thought he was going to attack the pony. But instead he stuck his head through the wooden fence and began enthusiastically licking her face. We all laughed, and Benjy turned around to look at us with something resembling a smile or a grin. We were hooked. It was all we needed: some sign that Benjy was a "special" dog. He was. He is! But I imagine that most dogs are special in this way. Each one has some quality that endears them in particular to their family.

Still, my disappointment continued when I found that Benjy did not obey me—indeed, did not even listen to me. When I asked him to come, he ignored me. Well, that was understandable. I was a complete stranger to him. What I found a bit more troubling was that when the two women who were taking care of him made a similar request, he looked similarly reluctant to obey. He was not sullen. He was not stubborn. He was not defiant. But he was clearly not eager to comply with commands. *This must be the reason he did not make the grade as a guide dog,* I thought. Naturally, guide dogs must be extremely obedient, and must subsume all their desires to those of their person. They could not have the faintest trait of disobedience. But we all *oohed* and *aahed* over Benjy's affectionate nature, as demonstrated by his licking the pony. Never, the women said, had they ever seen him chase another animal or even show the slightest inclination to do so. He could walk among sheep (of whom there are plenty in New Zealand) or near cats and one could be certain that he would not try to chase them or attack them or even bark at them. "He has not a trace of aggression in him, at least not that we have seen in these four months," one of the women told me. That remains the case to this day. I found this such an appealing quality that I could feel myself bending. The boys were jumping up and down—they knew this was "their" dog, and they absolutely had to have him. He belonged to us, they said. It was foreordained. Like a religious conversion, it had nothing to do with the head and all to do with the heart. There was little I could say. Leila melted and so did I. Benjy was to be ours.

And so the story continues. A few weeks later, officials from the Foundation of the Blind came to our house. They wanted to see for themselves the accommodations we had for Benjy. Did we have a fenced yard? We did not. On the other hand, our house was right on the beach, and the bay was completely protected. There were no cars on the beach. The whole area could only be accessed by a steep path going down a hill, through a forest of giant Pohutukawa trees, until the sea was reached. There are just ten houses on the beach, and in front of each is a strip of lawn, then sand, then the ocean. He would be completely safe here. The formalities were daunting, because he had been raised as a guide dog for the blind, and officially trained for some time by the organization. It was an elaborate procedure and anybody who took on a guide dog who either retired or did not fulfill the criteria was carefully vetted. Would we be able to give Benjy the life he deserved? They did not reveal what they were looking for, but we had to undergo several home visits to see if we were suitable adoptive parents. In the end we passed, I believe primarily because we said we did not plan to do anything that wouldn't include Benjy. If we went on holidays, so did Benjy. I would take him on all my shopping trips. He would hardly ever be left alone in the house.

If Benjy showed a great liking for the pony and for all children, he was a bit more standoffish with us, and with other adults. I did not initially notice his unusual qualities, but this is because, like an adopted child, it would take some time before he would, or could, be himself.

Indeed, Benjy's first days with us can be compared with the adoption of a child into a home. If a dog can love, then should we not expect a dog to follow the same process as humans—not to love blindly, but to acquire a loving nature by *being* loved over time?

It was clear that at first he did feel strange. This is not surprising. We were strangers, after all. He was good-natured about it, but he seemed removed, or at least remote. He was a bit indifferent and did not entirely trust us. Indeed, why should he? He did not know us. The American philosopher Glen Mazis, in *Dogs & Philosophy*, notes:

Our sense of location, of direction, and of orientation are more felt, in emotion, in the visceral depth of the body. This gives us our sense of "belonging"—an experience vital to both humans and dogs. When humans feel as if they don't belong, they're anxious, out of sorts, and stressed. Dogs, when they sense they are not where they feel at home, seem to experience the same feelings, as they whine or tear things up, or pace nervously.

Benjy didn't carry it too far, but he was, like an adopted child, very hesitant, tentative, not giving of himself at all, or only very slightly. Slowly, very slowly, he began to change as he felt more at home. As an American in New Zealand, I can understand. Sometimes I too feel alien, displaced, in exile. That is not just about the amount of time I have spent here; it is something else. I don't think Benjy's emotions on this issue are quite so complex. He cannot, of course, feel out of place in New Zealand compared to Europe or the States. He cannot make such comparisons and does not busy himself with other possibilities. That is a difference. Whether it is an advantage or not is unclear. But Benjy has been with us over two years now, and I can still see how each and every day, he gets closer to us. It has taken him this long to feel he truly belongs, that he is at home, that he is with family. Of course, he does not articulate these feelings, or "think" them, but he surely feels something like this. He may not contemplate the reason for the change, but the change is nonetheless real for him and for us.

There have been so many examples of Benjy's ordinary but extraordinary affection. *Ordinary but extraordinary affection.* I say it is ordinary because many, if not most, dogs display the same affection on a daily basis. I say it is extraordinary because it is powerfully visible. It is apparent to all who meet him that Benjy has a little something extra when it comes to loving. It shines right out of his face and hits you square in the heart; it cannot be missed. That said—and this book will be filled with accounts of Benjy's deep affection for all living beings—I must admit that this affection includes a somewhat strange quality. Once, we went to a commune in Northland on the north island of New Zealand. There were some 500 exquisite acres of native bush, streams, and mountains. Twelve Americans and a few Kiwis have lived there for years, and I have long wanted to visit and see for myself their vegan lifestyle (they eat no animals or animal products, even honey). They told me before we went that they loved all animals, but dogs in particular, and they felt very deprived, for their dog was living in Hawaii, where they spend half the year when not in New Zealand. So when we arrived, and Light, the commune's founder, came to greet us, he looked at Benjy with particular delight. I whispered to Benjy, "Look who's here!" (pretending that Benjy already knew him, though this was the first time he had ever seen him—I could see Benjy thinking, "I guess I forgot him,") and sent him off at a race to greet Light. Benjy was all over him, kissing and licking his face, giving him the tiny love-bites that he reserves for people he adores but has not seen for some time. You can imagine how happy Light was with this greeting.

But a critic of my thesis here might ask, how genuine was this love? Did Benjy respond to my words as almost an "order" to love? I thought about it for some time, because contained in this episode is the criticism one often hears of how dogs love indiscriminately. Dogs will love a serial murderer as much as they would love Gandhi. To them, a child abuser can still be a beloved figure. There is some truth to this "criticism" of dogs.

Some even go so far as to maintain that dogs are the ultimate hypocrites, bestowing their love on everyone equally because it is unreal, and is only a ploy to obtain what they want. I could spend an entire book refuting this argument. But here I only want to point out that even if there is some truth to the idea that dogs do not make the same distinctions as humans, and that they do love "indiscriminately," I still believe that this love is genuine. It is real. The fact that dogs do not make the kind of distinctions humans make does not mean their love is any less sincere. The feelings they have are heartfelt, even if they are not directed the way we would like to see them directed. They are no less extraordinary for being equitably distributed.

DOES BENJY LOVE even more when he is most happy? I think the answer is yes. For the two days we spent at the commune, he was in what could only be described as an ecstatic state. He had hiked in new mountains, something he loves. He had swum in many deep streams, another beloved activity. All twelve members of the commune petted him just about every minute we were there. He rolled on his back, showed his special happy grin, leapt in the air, kissed everyone who got close to him, and ran at breakneck speed around all the people there. He was a pure delight, and it was a delight to watch him. When they asked me what my next book was about, I had only to point to Benjy. They understood immediately!

He has utterly endeared himself to us. And he's caused me to wonder: What exactly is so unusual about how dogs love us? And what has it taught us?

CHAPTER 2

HOW DOGS LOVE

Dog love is unlike that of any other species—even that of humans. Conduct an experiment for yourself: Walk down the street and note every time you pass a dog on a leash, in a car, or any other circumstance. Look at the eyes of the dog. What do you see? The dog is looking back at you. Dogs make eye contact with strange humans and with strange dogs. Always. No other animal does this. Even humans don't do it: Most of the time, when we walk down the street, we avert our eyes from other people. Whether out of shyness, fear, or other reasons, we do not wish to make eye contact. And we certainly don't want to meet and greet a complete stranger. But dogs inevitably do. Even if one is tied up in front of a store, waiting for his human companion to come out, he will look at you. The look is not hostile, or fearful, or even evaluative. It is usually a friendly glance. This almost surreal friendliness is the primary characteristic of dogs. This quality of amiability, of sociability, of friendliness goes so far beyond what we normally see in other species that some people (count me among them) would describe it as a heightened capacity for love.

It is unusual in part because all species are to some extent tribal and exclusive. All make the us/them distinction, critical to survival. I am not the first person to have advanced the view that we are the only species that wages war and is guilty of genocide. No other animal species has ever engaged in anything even remotely like this. A pack of wolves might be vicious (from our point of view) but there is a productive endpoint of this behavior—to eat and survive. My friend Marc Bekoff reminded me that Jane Goodall, in fifty years of observation of chimpanzees, saw what looked like warfare only a *single* time. "They never thought of it," somebody might retort. Oh, a lack of intelligence, is it? Do we really wish to argue that it takes an act of imagination to wish to destroy every man, woman, and child of another human tribe? Strictly speaking, it does require thinking about the long-term future (one devoid of all Jews, say). No nonhuman animal is "capable" of this feat—agreed. But is this a quality for which we wish to be characterized as a species? Wiping out a whole group of people because they are identified as "other" is a dubious achievement. Surely no other animal has ever even contemplated this. Dogs, alone among animals, do not seem to make the us/them distinction. Some even believe they are human. Or they think *we* are dogs. In any event, they do not make the distinction that all other animals make.

But what about animals that dogs seem primed to view as prey, such as squirrels and rabbits? It takes considerable effort to get your dog to give up the chase instinct in tempting situations. Ask the lion tamer. Or think of terriers. They are loving to their families, but hundreds of years of breeding for the kill instinct makes it very hard to convince them that a small moving rodent is now part of the extended family. "Are you crazy, boss?" is the likely response. But it can be done. I wrote a book about how different animals, when raised together, form friendships and even loving bonds (*Raising the Peaceable Kingdom*) and I even have a photograph in that book of my Bengal (half-wild) cat eating from the same dish as our pet rat. That was hard for our cat: To overcome a cat's inborn urges is no small matter. Getting the dog to comply with our wishes was easy in comparison because the instinct to chase and eat prey has to compete with the instinct to please us, to share in the love.

Dogs are eager to see and meet other dogs and people; they are eager to play with other dogs and people; they like to walk with other dogs and people; they like to sleep with other dogs and people; they are eager to express affection to other dogs and people. But even more important, or more astonishing, they are willing to do these and other activities not only with people and dogs, but with many other animals as well. Instantly. Unmistakably. Inevitably. Can we say this of any other species? No.

Consider the cat. Cats meeting other cats are instantly, unmistakably, and inevitably wary and often unfriendly. They don't trust one another. They assume, usually correctly, that unplanned and unwanted encounters are likely to end up as a disaster, as a fight. Cats avoid such encounters. All the big cats—in fact, the entire feline family—are notoriously shy or suspicious of one another, let alone an animal from a different species. Whether through breeding or intense training, our domestic cats have shed this trait to some extent. They can learn to live peaceably and even on friendly terms with other cats, dogs, birds, and even rats. Our three cats hunt wild rats, but have somehow learned that our "pet" rats belong to a different, protected category. There is the remarkable story of Chris the lion, which millions of people have watched on YouTube. Two Australians kept a pet lion, Christian, in London for several months in 1969, and then handed him over to George Adamson, wildlife conservationist and author of the book *Born Free*, to be released in Kenya. A year later they visited Kenya and, amazingly, found their now wild lion. Christian greeted them with boisterous delight and friendly remembrance, jumping onto his hind legs and literally throwing his arms around his former caretaker's neck. We would not be surprised had Christian been a dog. But we do not expect any of the big cats to display affection, and especially so after a long absence. Dogs, of course, never forget their love. Tigers and lions, on the other hand, frequently kill their trainers. One of the most famous episodes happened in 2003, when a seven-year-old, 600-pound white tiger grabbed Roy Horn (of Siegfried & Roy) by the neck, dragging him offstage in front of a horrified audience—some of whom thought it was part of the act—and nearly killing him in the process. Here was an animal born and raised in captivity being "handled" by an experienced trainer (it was Horn's fifty-ninth birthday the day he was mauled).

It is not only the cat family; bears are notoriously solitary and dangerous, both to one another and to other species. Dancing bears (bears, particularly in India, forced to dance for the amusement of a human audience) are a disgrace to any society that indulges in them and the process that brings these bears to their captive state is not worth repeating. The fantasy that we can befriend a wild bear is best kept to children's books and movies. When the fantasy is indulged, tragedy is the result. Just how dangerous are grizzlies? Very. You have only to watch the mesmerizing Werner Herzog film *Grizzly Man* to see how dangerous certain idealistic but naïve attitudes toward these animals can become. Timothy Treadwell spent twelve years with the grizzlies in Alaska, shooting 100 hours of extraordinary footage at close range. His own Web site (www.grizzlypeople.com) warned against getting closer than 100 yards to a bear, yet he was often right up in their faces, even touching their noses with his finger. Bad idea. In 2003, just minutes after filming, he and his girlfriend were attacked by a male grizzly with whom he was not familiar. Both were killed.

Notice too that animals that live in large colonies tend to be herbivorous and harmless. The more dangerous an animal is, or is perceived to be, the less likely other animals will find it wise to stay in the vicinity. Elephants are an exception in that we saw them, as did most noncarnivorous animals, as benign. Or rather, they were considered so until recently. Possibly because we have destroyed so much of their habitat and their family structure by killing matriarchs for their tusks, in the recent past elephants have gone on rampages in Uganda, South Africa, and Bangladesh, killing rhinos, humans, and other elephants. Orcas (killer whales) rarely show aggression in the wild unless they are feeding, but in captivity can be extremely dangerous to themselves, other cetaceans, and even their "handlers" (there are some two dozen incidents known). Gorillas rarely engage in any form of violence. The silverback male may, upon meeting another silverback, engage in mock charges and vocal terror, but rarely do the two wind up hurting each other, and it is extremely rare that gorillas hurt any other animal (they are pretty much obligate vegetarians except for the occasional insect). King Kong notwithstanding, gorillas don't usually hurt humans.

But none of these animals, whether harmless or dangerous, carnivore or herbivore, solitary or sociable, are ever friendly in the wild with members of a different species, or even with a strange member of their own species. Such encounters can be deadly.

DOGS REMAIN THE exception. They are like no other animal, certainly no other wild animal. How can we not be enchanted? There you are, watching a wolf romp on your bed, lick the face of your child, and chomp gently on your arm. You feel safe, protected, and loving, and the dog feels the same way about you. And dogs have overcome the wariness and distrust of the wild wolf. Wild wolves, upon meeting other (strange) wolves, are unpredictable. David Mech, the world's leading authority on wolves, has pointed out how many wolves kill other wolves. Dogs, on the other hand, assume correctly that strange dogs want to be their friend. Dogs are the Bill Clintons of the animal world: They have friends everywhere and take it for granted that each and every dog they meet will be happy to play and romp and show affection.

Sometimes when people admit that dogs live in the moment, they add a caveat: Yes, but that's because they are always supremely self-involved and self-interested. Dogs are narcissists, some people argue; it is always about "What's in it for me? Have you something I can eat?" I disagree. Dogs are constantly evaluating what their senses bring them, yes. They are supremely observant, seeing and hearing everything, evaluating how much of an impact it will have on them. They're always hungry, after all! We could regard this as a narcissistic trait, but are they really "lost in their own little worlds" in the human sense? And of course, a true narcissist is not interested in your pain. Most dogs will do just about anything to assuage your suffering, even to the point of putting themselves at a disadvantage or even in danger. I would not call this narcissism.

Dog love is not just a question of coming from a sociable species. Many other sociable animals may be able to engage in peaceful coexistence, but they rarely show the unbridled enthusiasm for a member of their own species that dogs routinely do. Not only do dogs love to see dogs they already know ("It's you! It's you! I can't believe it's you!"), they are curious about each and every dog they *don't* know as well ("Tell me your life story!" is their default greeting). It never fails to amaze humans that a giant Saint Bernard greets the tiniest Chihuahua as an MOT (a "member of the tribe," as my parents used to say when they met another Jew), never doubting for a second that they are a dog too. Cows can be peaceful with other cows, but they are unlikely to express joy at seeing another cow arrive in a pasture, or at least will not demonstrate the wild enthusiasm we see daily in dog upon dog encounters. Sheep, goats, chickens, and other domesticated animals on a farm react in the same passive way. They are capable of forming interspecies friendships and of tolerating a wide variety of individuals, but we have to wonder, are they really into it? Are they moved to joy every time they see a member of their own species? I think only dogs display the unbridled pleasure we call *joie de vivre* routinely.

Dogs carry this same enthusiasm for meeting a member of their own species to meeting a member of another species, namely us. What is everybody's favorite dog trait? Easy. You leave the house in the morning, but you have forgotten something, so you return a minute later, and your dog is overjoyed to see you. It is not that he is happy because he thinks you realized your error in leaving him in the first place; it is just that he is happy to see you *all the time*. It doesn't matter if he just saw you five minutes ago. Each and every time you return from the shortest of visits, there he is, waiting to greet you. Benjy has now taken to sitting like a lion on the path that leads to our house, waiting for me to return. When he catches sight of me at a distance, he races toward me, his ears flapping, at breakneck speed and then greet me as if I had just returned from the dead. No other animal, not even other humans, are as engaged in the business of making friends with strange humans as dogs. (I keep typing *gods* by mistake—I suppose there is some truth to the error. Humans like to imagine that God will love them when nobody else will. The truth is, however, that only dogs will love you when nobody else will.) Why? For one thing, it gives them enormous pleasure. But then we must ask why meeting a complete stranger should give anyone pleasure, and unbridled pleasure at that? It is not just that they anticipate a play session (they often do and often get it), because sometimes they are content to merely make eye contact, lick a hand, or get a pat or some other fleeting indication of friendship and affection. Somehow this reassures them of something essential to the nature of a dog.

While a few species of animals play as adults (marine mammals such as dolphins, bonobos, and river otters come to mind, and I have even seen some adult goats play), perhaps no other species plays so compulsively and at such short notice: "Hi, I'm Benjy. Wanna play?" Strange dog: "Sure. How about now?" And just as with the demonstration of affection, they often direct their play requests to a member of another species. Most dogs will play with most humans even if they have never met them before. If it is throwing a stick, they don't seem to care who throws the stick. If you want to chase them in play, they don't care if you are fast or slow, young or old. No animal in the wild plays, except under very unusual circumstances, with a member of another species. It probably never even occurs to them, any more than it would occur to us to want to play with an ant. This is one reason people were so astonished to see a video of a polar bear (who looked like an adult, but could easily have been a juvenile) playing with a sled dog. It was not odd to note that the dog would want to play with a bear, but scientists found it puzzling that a fully grown bear would take pleasure in activity so unlikely to lead to any immediate benefit for the bear. The video is everywhere on the Net—millions watched the play. There were two huskies, both chained. They belonged to a German photographer who was shooting in the Canadian north. A thousand-pound polar bear approached the dogs. Evidently he had not eaten in four months. The photographer was convinced it was curtains for his two dogs. Instead, one of the huskies made a play bow, and the bear seemed to understand that the dog wished to play, not fight. Play they did, for hours. In fact, for the entire week, every evening the polar bear would return to play with the dogs. In one scene, the husky actually turns his unprotected neck to the fangs of the polar bear, who only pretends to bite him. Nothing like this had ever been seen before. Are even wild bears susceptible to the charms of dogs?

Dogs seem to have an infinite variety of ways to express their affection and their love and their friendliness. They make it extremely easy on us to understand their intentions. It begins with the wag of their tails. Scientists tell us that there are many variations, but I see just different gradations of enthusiasm in Benjy. (Watch your dog sleep: The tail will never wag, because *you* are not there to observe it; in other words, tail wagging is just another means for dogs to communicate with us—and what they mostly want to communicate is pleasure.) Granted, an animal wagging its tail is not invariably a sign of friendship: As we know, cats wag their tails when they are angry or feeling aggressive. But when a dog wags his or her tail, we note also the expression in the eyes; there is a brightness and a sweetness that is very easy for us to read. There is the dog smile that I have already mentioned, because it seems unique to humans and dogs (chimps grin, but it usually indicates uncertainty rather than friendliness). There is a posture too that we find easy to read: Dogs move their whole bodies in joy. The body is not held stiff, but loosely. Then there are vocal signs of pleasure: whining, a soft bark, and a kind of moan of anticipated pleasure. I am not suggesting that other animals don't have equally varied physical characteristics of expressions of pleasure, but we are certainly less adept at reading them, and few animals, even among our domestic companions, have taken the trouble to make it easy for us (the purr may be an exception). Did dogs do this deliberately? Or have we been together so long that we have become adept at reading one another? Dogs read us just as easily—in fact, probably far more easily than we read them. Witness the many accounts of dogs who try to cheer us up when we are down, who know to avoid us when we are angry, who can anticipate an imminent walk even before we have any conscious intention of going on one. They are masters of knowing the time. When we get up for the afternoon walk, they know by the way we get up that we are not simply getting a glass of water. They have anticipated that we have made the decision to take them for a walk, and they are almost invariably correct.

There is not a day when Benjy does not teach me something about the nature of dogs, the nature of humans, and the nature of love. We were Benjy's fifth family in less than two years. For a young dog, that is a lot of change. I suspect all that moving was less than reassuring to him, so it took a while for him to embrace us. When Benjy came to us he was polite, but he had an expression of sadness about him that seemed indelible. He must have learned that nothing was permanent in his life. He disappointed people. Whether he knew it or not, this was the reason he was moved from one home to another. He was always expected to do something he either could not or would not perform. He could not or would not be a guide dog for the blind. He could not or would not be a seizure-alert dog. He wanted, I believe, to be a simple family dog, and when he finally got to our family, that is exactly what we wanted of him too. But he could not know that at first, and I believe he was waiting to be re-homed for a sixth time. It took time for him to recognize that he was not leaving—that this was indeed his home and his family. As he came to realize this, his behavior gradually changed. Today, he is more confident. He shows greater happiness. He smiles more readily. Not everyone believes that dogs smile, but I definitely do, and Benjy definitely does: mouth open, tongue out, slight panting—it is just so obvious. I wish I could introduce him to a skeptic and say, "Look, his eyes are smiling too!" The skeptic would look and say (I hope), "Yes, it does appear that his eyes are expressing happiness."

I can't help but wonder what Benjy felt when he joined our family. Did he believe that he was somehow lacking? Did he believe he was a failure, or even, God forbid, that he was simply not lovable? It is, I know, difficult to enter the mind of a dog. But I know for sure that Benjy was not a contented dog when he came to us. You know an unhappy dog when you see one. Remember, he was so sad that his previous family put him on Prozac. To some people—me, for instance—this is absurd. But it is not that uncommon, and many dog behaviorists are in favor of it. Nicolas Dodman, for example, in his 1997 book *The Dog Who Loved Too Much*, tells stories of anxious dogs who needed drugs, or so he claims. I remain very skeptical, and now Eli Lilly has a pet-focused unit, the first release of which was Reconcile—basically doggie Prozac. My main concern is this: Dogs need love, and when they're not getting it, Prozac is not the answer.

Whatever the solution, the problem was evident: Benjy was depressed. He was lethargic, his tail rarely wagged, he kept his head down, and he rarely smiled. Benjy may well have thought this home was going to be temporary too, and so why would he emerge from his sadness if he was to lose this family as well? But slowly, very slowly, Benjy began to change. He became more playful by the day. The look on his face altered too. His eyes began to shine more brightly. His demeanor altered. He held himself higher; his head no longer drooped. Every day there was some change in Benjy. We all noticed it: evidence that he was beginning to believe he was with us to stay.

For a long time he would take every opportunity to leave the house and wander the beach by himself. We live on a beach where there are no cars; the only access is a long path down a steep hill, so Benjy is not in danger, and we were reluctant to confine him when he appeared so eager to be outside. It seemed natural, but now I know it was not natural. Benjy wanted to be outdoors because he did not feel close to us.

By contrast, today, two years later, Benjy rarely goes outdoors on his own, even though all the doors are wide open. He waits for me to go. He lies beside me as I write this book. It took this long, though, for me to see that unmistakable look in his eye, the one that says, "I am glad I am with you."

Nor is it just our family. Now, whenever he meets somebody he has known well in the past (even the people with whom he lived when he was depressed—figure that out!), he cannot be restrained. He knows the command "down," but he cannot—literally cannot—obey such a command, as he is under a far more powerful force: He must demonstrate his love and recognition. So up he bounds, kissing the person full on the face over and over. Benjy is certainly the best example I know of how love and stability bring happiness into the life of a dog. I get to watch daily as he becomes ever more confident in this love. A dog needs this every bit as much as a child.

But I would be remiss if I did not also mention that dogs respond in the opposite direction to loss and deprivation of a loved one or ones—again, very much like a child would. I am glad to have not experienced this directly, but I have read many convincing accounts. Perhaps the most moving of all is *Nikki: The Story of a Dog*, by the Hungarian writer Tibor Déry, first published in 1956 and since considered by many a masterpiece, a classic of both understated political writing (the book was banned in Communist Hungary) and of unsentimental but affectionate observation of a dog, Nikki. The "master" in the book is sent to prison for no known offense, and Nikki is left with his master's bereaved, lonely, and increasingly impoverished wife. Nikki spends most of his time longing for his human friend. The author convincingly and heartbreakingly writes about how desperate the dog was for human warmth and companionship:

> *What gave her most pain was the terrier's silence, the muteness of her whole body. The bitch neither cried, nor argued, nor protested, nor demanded explanations; she*

> *simply resigned herself to her fate in silence. This silence, which resembled the ultimate silence of a prisoner broken in body and soul, was, for Mrs. Ancsa, like a violent protest at the nature of existence itself. At no time did she feel so keenly the grievous tragedy of the defenseless as when, her shopping-basket on her arm, she turned on the threshold to look at the bitch, who lay mute and motionless in the middle of the hall, staring fixedly, her head between her forepaws.*

Benjy's most characteristic pose is exactly the same: putting his head between his paws, but instead looking at us with utter contentment, letting us know how right his world has become.

HOW DO DOGS express happiness? Apart from the smile, there are gestures that all dogs seem to have in common, some obvious, some quite subtle. Benjy loves to roll on the grass or the sand with his paws in the air, with that peculiar goofy look on his face that says he is enjoying himself immensely. Sometimes as he does this, he talks to himself (that's what it sounds like to me—a mumbling mixture of soft growls, yips, and minor barks). At other times he puts his head down on the sand between his paws and raises his backside with his tail wagging furiously.

When it comes to play, perhaps no other animal except the human animal has been more studied than dogs. It may seem that dogs merely have more ways to signal play than other animals. Or we may be more tuned-in to dog play, because we so often participate. If I feint a sudden rapid move in Benjy's direction, he immediately assumes I intend to play with him, and he feints back. He becomes wildly exuberant, but only for a few moments. It is soon apparent to him that I don't know the proper rules. I feel like somebody visiting a foreign country and being invited to join a complicated ball game. I am willing to try, but appear faintly ridiculous because I just don't get it.

One crucial distinction between dogs and other species is the fact that dogs often *assume* an affectionate response from strangers, both of the human and dog variety. In the beginning, Benjy treated the cats as if they were dogs. But, being cats, they refused to engage. He made a play bow; they ignored him. He rushed them; they stood their ground. Then he became insistent, more exuberant, and rushed straight at them with great speed. They weren't taking any chances and they weren't playing. They left in disgust. Benjy looked disappointed, like a kid left alone in a sandbox. He only meant to play, not intimidate, and certainly didn't want to be left to play by himself.

Sometimes we will be walking down the street when suddenly Benjy becomes possessed. He sees somebody, and I imagine he must know and recognize the person, for he rushes over and gives his full greeting ritual. This is the one thing I have not been able to train out of him or extinguish. He leaps up and plants kisses on the person's face, wrapping his front legs around him or her in what resembles nothing so much as a hug. It often turns out that he does *not* know the person—but he must know *something* about them, for not once does he make a mistake and chase somebody who does not like (make that *adore*) dogs. Invariably they are thrilled; they hug him back and make a great fuss over him. He looks over to me as if to say, "See, I know how to pick them." And he does indeed. I have given this a great deal of thought, and I believe that Benjy notices slight and subtle physical gestures and clues that the rest of us ignore. He sees a smile or an appreciative glance in his direction and he reads it correctly: "This man or woman will welcome my fulsome attention."

BENJY IS STARTING to have better luck with our cats. Benjy came to us when he was two; the cats were much older and had been with us for more than six years. Their domain extended beyond our house (which they ruled utterly, of course) to include the beach in front of our house, as well as anyone who should happen to be picnicking, swimming, or simply walking along the shore of that beach. So I anticipated trouble when Benjy joined our household. But I forgot that we are not the only species able to read the body language of dogs. Cats do it just as well. It is not that they misread a wagging tail; they just sometimes choose to ignore it or treat it as a declaration of war because, well, they are cats, and they can do whatever they like.

I was wrong to anticipate trouble between Benjy and our cats. The cats did protest in the beginning, but as Benjy continued to wallow in humility, rolling over to expose his most vulnerable parts, crawling on his stomach, oozing deference and subservience, the cats slowly began to relent. Just to make sure he got the picture of how serious the evening walk was for all three of them, the first time he joined us on the beach, two of the cats walked over to him and nonchalantly gave him smacks in the face with their paws, while the third looked on with approval. He bowed his head and made not a peep. Then we were able to continue the walk. But he is an exuberant Lab and he likes to walk by racing in circles. This nonplussed the cats. I thought Benjy was going to get smacked again, but I was wrong: Megala, the ringleader of the three cats, put a stop to

Benjy's boisterous activities by racing up to him sideways. Megala was actually galloping, and it stopped Benjy in his tracks. Benjy thought it was pretty funny (so did we), but he backed off just in case. Over the next few months, this became a routine game: Benjy runs up to the cats in a faux attack (how they know it is faux is a mystery, but they definitely know it is not for real), then turns around while one or two (or sometimes all three) cats chase him back, with Megala always moving at a great sideways speed. (I think Megala believes he looks bigger that way—and come to think of it, he did seem bigger.) It's not just fun; it's enchanting. What allows our three cats to understand Benjy and his benign intentions so well? What interspecies communication are they engaging in that has *not* been hardwired into them by evolution? How are they able to make this transition so effortlessly? It would not, could not, take place in the wild. Is this play an indication that there is something to be said for domestication, and something to be gained by animals living with humans, even as glorified slaves?

GENERALLY SPEAKING, WITH their companion humans, dogs seek every possible opening for affection. Manu, our eight-year-old son, still has the habit of wanting one of us to lie with him when he goes to sleep. But we have discovered that can be Leila, Benjy, or me. We discovered it by accident. Benjy was walking by the room where Manu sleeps when he heard words he may not have understood, but whose music was all too clear to him. Affection was being doled out, and wherever there was love, there had to be Benjy. So in he wandered, leapt up onto the bed, stretched himself out fully, placed a large leg and paw around Manu. Within seconds Benjy was snoring (he has the amazing ability to fall asleep at any time, anywhere), his leg securely around Manu. Manu was at first startled, then delighted, then dismissive: *You can go now, Mom—I have my cuddle-buddy.* And so that has now become our new bedtime ritual: Mom or I begin and then Benjy takes over. All are happy. (Except for the strict dog-trainers: "How can you let Benjy believe he is leader of the pack? He must not be allowed to sleep on your bed.") Even less strict trainers would find it an odd arrangement, not one conducive to teaching a dog his place. But what is Benjy's place? If you believe in hierarchy and how it must always be honored, then perhaps you will find these trainers correct. But if you believe as I do—that Benjy represents a certain kind of mutation, and that he is pure love—then this is merely one more expression of that most delightful of genetic mishaps: a creature entirely benign and wishing only to express the immense love he feels inside. I have no complaints.

What is this love? For many of you reading this, it is obvious. You know exactly what this love is, because you have both been on the receiving end of it from your dog and bestowed it on your dog. It is love, plain and simple; that's all, you say. Yes, I agree with this. But if you stop to think about it, then you will have to agree that this love is really quite a remarkable phenomenon. It is, in fact, one of the most remarkable phenomena in our universe. How does it come about that two members of completely unrelated species can feel such deep love for each other? There is nothing else quite like it in the natural world.

Benjy is not unique. Even though I like to think that his love is so deep and so boundless that it reaches proportions that are almost miraculous, in reality I know perfectly well that many of you have had similar experiences with your dogs. You have had dogs who risked their lives for you or who loved you beyond all measure. You know of this love because you have direct experience of it, just as I do. Like me, you have looked into your dog's eyes—and he or she has stared back, mirroring your love. We can read love in a dog's eyes. That in itself is something of a miracle. We find it much harder to read the eyes of our cats, our parrots, or our horses. We cannot know for sure what they are feeling simply by looking them deeply in the eyes.

I put my philosophy of love—*If I love you, then you will love me back*—to the test once with a wild elephant. I was dead wrong: I was not seeing compassion or friendship in this giant female elephant's eyes, but rage that I was intruding upon her personal space in the middle of the South Indian jungle. I had no right to be there, and she attempted to teach me manners by almost trampling me to death. I escaped by a miracle and learned (to my chagrin) that I had absolutely no mutual bond with wild elephants. Probably nobody else does either. Elephants love one another, of that I have not the slightest doubt. Nor do I doubt that the love they feel is every bit as real, deep, and interesting as human love. But it is confined to other elephants and does not ever truly cross the species barrier (at least in the wild—in captivity, there can be exceptions). If that elephant spared my life, she did so not out of love or concern or even affection, but purely by accident. Had she killed me, it would not have involved any strong emotion on her part, either. I was the fly on her back about to be swatted into nonexistence, no more.

CAN I BE so certain of this assertion, this unique bond of love between dogs and humans? I think I can. I have lived with many animals over my nearly seventy years. I have made a particular point of studying the positive emotions between other animals and humans, partly because for so long this was a neglected, almost orphan, field (in the past ten years this has changed). I have written much about these emotions. I have loved spending time with chickens and many other birds, with rats, rabbits, hamsters, and cats. I had ducks as a child that would have followed me to school had my parents permitted. As it was, they waited for me to come home from school so we could be together. I have studied farm animals and seen the strong bonds that can develop between pigs, in particular, and humans. Even cows and sheep can learn to like our company. Many people adore goats. I have talked to many "horse people" and seen their deep connection to their horses. Like many of you, I have especially enjoyed sharing my life with cats. I have witnessed many moments of mutual affection—sometimes the affection was so deep that it seemed to pass over into love. But this was mainly on the part of the person.

Ilan's rats, Kia, Ora, and Rani, are very affectionate. One was lost for three days in a forest near a gas station. When we returned and found her in a hedge, she went racing over to Ilan in obvious relief to be found. But I am a little reluctant to call what the rats felt for Ilan "love," even though Ilan has no such hesitation. Perhaps it is because I am not the direct recipient of the rats' affections that I waver a bit here. It could also be due to the prejudice so many of us share when we refuse to acknowledge complex emotions such as love in an animal as small as a rat. Or perhaps it is the bias we inherit about rats, seeing them—falsely, I believe—as lesser animals who bring disease in their wake. Domestic rats, I know, make delightful companions. Certainly what they felt for Ilan was similar to what they felt for their sister rats, and what, after all, is wrong with calling that love?

WHEN I ASK adults who live with horses, and cherish them deeply, whether their horses love them back, most have been honest enough to say no. The horses offer affection, yes, sometimes even gratitude. But love, love as we know it between one person and another? No.

Of course, dogs make it easy for us. Perhaps they don't trust us to be able to read the signs correctly, so they invented the wagging tail. Dogs don't deceive us: You will not get bitten while patting a dog with a wagging tail (unless it is held stiff and is wagging slowly and the dog looks angry), and few humans would pat a growling dog. But the eyes on a dog are special, because their expressions are so similar to ours. *Our* eyes, *human* eyes, also can express love and pleasure. We too smile, and it is an invitation. (When other primates closely related to us pull back their lips in a grimace, it is not one of pleasure—and I have never seen a cat smile except in paintings.) Unlike dogs, we can deceive, and we do deceive. "One may smile, and smile, and be a villain," said Shakespeare in *Hamlet*.

As I was thinking about this, Benjy put his head on my lap, looked up at me, and sighed as our eyes locked. There it was again, that look in the eyes. I tried it with the most affectionate of my three cats, Megala. I looked him deep in the eyes and he didn't like it one bit. He lifted his paw in warning—I was about to get scratched for my effort. I stopped. He looked relieved that he did not need to teach me manners, and walked away in a huff. Cats do not look into one another's eyes unless they are spoiling for a fight. It is a dare, an aggressive act. I don't think Megala cared whether I knew the proper feline etiquette or not—he wasn't taking any shit from me. In fact, he never does.

None of my cats do. Any mistake is swiftly punished. They are like punitive parents. They don't know you are more likely to be persuaded with kind words than a slap on the wrist. Of course, all this depends on their mood too. If they are in the mood for it, they can be ever so gentle, relaxed, and friendly. But always on their terms. If they are not in the mood, they cannot be coerced or even coaxed into a friendly exchange. They just walk away.

Not so Benjy or his peers. They are *always* in the mood for a friendly exchange. *Always* in the mood for pats, expressions of approval, words of encouragement. They love to hear us say how much we like them. They may not know the meaning of the many endearments we use for them, but they certainly recognize the music in our voices. The tone appeals to them and appeals to them endlessly. "More" is their constant refrain. You can make your dog achieve something resembling ecstasy by putting more and more enthusiasm into your voice until he is positively bursting with good feelings. And you can do this over and over. Nothing is so urgent for a dog that he will not stop instantly for a display of love.

Perhaps you don't need to be convinced. You live or have lived with dogs and know exactly what I am talking about. OK, but here is the thing: Have you ever stopped to consider *why* this should be so? I am not the first to notice it, obviously, but surely it is a matter of some considerable importance that this extraordinary, almost preternatural ability to love exists in two such different species. You might expect it to be found in two species that are far more similar, such as humans and chimpanzees or any of the other great apes, gorillas, bonobos, or orangutans. But it doesn't. There is that touching moment that Dian Fossey has written about, when one of the gorillas finally trusts her enough to reach out and touch her hand (made famous by the lovely image of Sigourney Weaver in the film version of Dian Fossey's life, *Gorillas in the Mist*). It thrills us partly because it is so unexpected, but also because we do not really believe that love can exist between an entirely wild animal and a person. And we are probably right. This is not love; it is trust, a certain degree of getting to know the "other." It is absolutely compelling, and rightly so. But it is not love of the kind we have with our dogs. The juvenile gorilla does not rise and follow Dian Fossey back to camp to live with her as a faithful and loving companion for the rest of their lives. We don't expect that from other species; it has never happened and probably never will.

So why are we not amazed that it happens daily with dogs? After all, if a wild wolf were to walk out of the forest and begin to wag his tail at the first human he saw, then walk home with the person and settle down comfortably in front of a fire, wagging his tail as the cats walked by, it would be front-page news in every newspaper on the planet. But we are simply familiar with such behavior in dogs, and once we are used to something, no matter how astonishing, it loses its power to keep us enthralled.

The other answer is that we regard this as a human accomplishment: The dog is *the* domesticated animal *par excellence*. We have created, out of the wolf, the dog—that is, we have done precisely what I said was so remarkable. Were this to happen today, it *would* make front-page news across the world: *Extra, extra, read all about it! Wolf domesticated! Dog created out of wolf!*

Why is the feeling about the domestication of the dog so different than when other animals were domesticated? "It's because they give us unconditional love"—I hear that a lot. I believe it. But if you stop to consider it, the remarkable thing is that we give it back: When we love our dogs, we love them unconditionally. It is a powerful feeling. People often say that the strongest love they feel is for their family and for their dog. So again, I have to ask if this strange coincidence is not more than that. Is it not possible that over the many thousands of years dogs and humans have been together we have taught one another this special kind of love, one that we reserve for only a few other members of our own species, and for one single other species? I believe it is.

CHAPTER 3

DOGS MADE US HUMAN

We are a storytelling species, and in this, perhaps, lies that ever-sought chimera: human uniqueness. As far as we know, only we tell our children stories to put them to bed; only we tell one another our very own stories, our own biographies, just because we like to know where we came from.

Some of the stories we tell are about our children and our family members: stories of belonging, stories of beloved moments, stories we like to repeat over and over. And when these stories are about animals, they tend to be about the dogs in our lives. We give our dogs biographies. We know them to be the subject of a life, to use the terms of the animal rights philosopher, Tom Regan. They do not have stories to tell, but we have stories about them. We know of their infancies, sometimes their births, and their days of childhood, their glory days, and, with heartbreaking frequency, we recount their declines and their deaths.

Of course, all animals have such stories. It is just that the stories are not always attached to them by humans. And with nobody to listen to their stories, to give them currency, or to recount them to others, they disappear. But that does not make the life that was led any less real, any less dramatic or poignant. Dogs alone escape this anonymity, and their stories live on long after they've passed from this world. People in different countries tell them to their children in illustration of the wonderful things dogs do for us. We still know the stories about dogs from the Greek epics (especially Argos in Homer's *Odyssey*).

People in India tell their children about the faithful dog in the Indian epic the Mahabharata, the core of which was written well before the Christian era. As the story goes, the great and righteous Yudhishtira abdicated the throne in favor of a great-nephew. Then he, along with his four brothers and queen Draupadi, set off on a pilgrimage, which they knew would be their final journey. As they moved north toward the Himalayas, a dog attached himself to the royal procession. One by one, great warriors (including Krishna) and the queen perished, leaving only Yudhishtira and the faithful dog to continue on the journey alone. Yudhishtira was lonely and numb with grief, but he pressed on. Suddenly, a glow of light appeared, and out of the light came the god Indra in his golden chariot. Indra invited Yudhishtira to board the chariot, since it would carry him straight to heaven. As Yudhishtira prepared to mount the chariot with the dog, Indra laughed and told him that there was no place in heaven for a dog. Yudhishtira shook his head sadly and said he would not mount. "This dog," he explained, "has shared all my troubles. He is devoted to me. I will not abandon him, not even for heaven. I have a duty to this dog, and not all the joy of heaven will persuade me to leave him." Indra tried to persuade him and urged him to "give up this obsession for a dog." Yudhishtira was firm: "I can't give up this dog for the sake of my happiness." Indra insisted. He reminded Yudhishtira that he'd given up his brothers and his wife. "No," said Yudhishtira. "I only gave them up when they were dead. This dog is alive. He stays with me."

Yudhishtira turned away, but as he did so, the dog began to change form. None other than the god of justice himself—Dharma, Yudhishtira's celestial father—stood before him. All along, it was Dharma who had kept pace with his son, all the way until the final test. Yudhishtira had mirrored the dog's pure devotion, and for this he was rewarded with his place in heaven.

THESE ANCIENT STORIES tell us that dogs have been crucial to us for a long time. But how did that come to happen? The process was unconnected to the domestication of other species—pigs, cows, and sheep, for example, all of whom were introduced in the last 8,000 years, long after the wolf turned into a dog. This early date (whether we accept 100,000 years, 40,000 years, or even just 15,000 years) makes the domestication of the dog different and more significant to humans than that of any other animal. There are no ancient stories about other domesticated species in world epics, as there are about dogs. Even if we accept the latest date for dog domestication, 15,000 years, if we think of that time in terms of other events in human history, we realize just how significant it is. There was no such thing as agriculture; there were no seeds we planted and farmed for eventual human consumption. Humans were hunter-gatherers (although the term is slightly confusing; evidence seems to show now that we did considerably more gathering than hunting). Our entire food source was to be found in the wild and most of it came from plants: wild fruits, berries, edible leaves, bark, roots, and tubers.

We did do some hunting, but we were probably more hunted than hunter. The big cats could and did easily take us as prey. We had weapons 15,000 years ago, if not terribly sophisticated ones (the bow and arrow, for example, was to come later), and fire too. But we felt vulnerable, and it was more than a feeling—we *were* vulnerable. You know how little children like to fantasize that they have an all-powerful animal who can protect them from all harm? (The charming dog movie *Bolt* plays with this theme, and C. S. Lewis makes a lion into God in *Narnia*.) I believe human ancestors had this fantasy too, and then discovered that it could be more than a fantasy—humans *could* find an animal willing to protect them. This is, after all, the standard view of how dogs came to be domesticated. Many scientists—in particular, Juliet Clutton-Brock—have imagined the scenario.

According to one typical version, men were hunting in the forest and may have frightened a female wolf away from her den, inside of which were young cubs. Most cubs ran from the men, but one, bolder and more curious than the others, did not. He approached. The men were touched and did not harm him. They brought him back to the campsite. Soon their human children and the cub were playing happily together. Or the scenario is slightly different: A mother wolf is killed and her cubs, too young to care for themselves, disperse after searching for her. One comes upon a human encampment and is seen by children. (Whatever scenario one imagines, children tend to be part of it, for by their very nature they share traits with the cubs: playfulness, curiosity, and a certain reluctance to harm another small, vulnerable creature.) The fact is that at some point in our prehistory, we were bound to run into wolf cubs. It would not (and it still does not today) take much to tame a wolf cub. Once such a cub grew larger and stronger, it became clear to humans that they had a potential ally—in fact, more than an ally: The animal would have been a reliable and remarkable defender. (At least some of the time; I will point out that wolves kept as pets today are very dangerous, as are wolf hybrids, so they must have been equally dangerous then as well.)

Being territorial, the wolf—now grown large and strong—would consider the human encampment his territory as well. Adult wolves don't bark, but certainly the wolf would guard the encampment, and it would be a rare large cat that would approach such a place guarded by several large wolves. Moreover, while humans were not great hunters, they were certainly persistent and intelligent hunters. Since their new wolf companions often tagged along with them, they were soon able to see that wolves had skills humans did not possess: They were much faster, had four times better hearing (or even far more, depending on what you believe), and an infinitely better sense of smell. A wolf could chase an animal into a tree and his human companions could come along and make the final kill. They would share the flesh. As I say, these are only likely scenarios. We have, obviously, no written record, and no pictorial ones either.

"But these are wolves," you might object. How did we get a dog from a wolf? Again, we are forced to rely purely on our imagination for an answer. But in this case, what we imagine is also what actually happened much later in the domestication of other animals. Therefore we are not just guessing, but reconstructing from known cases. A domesticated animal is essentially an animal that is entirely under our control. That is, we do not expect this animal to harm us or even challenge us. We do expect, however, that this animal will run away from us; that is why all later domesticated animals must be confined in one way or another. We do not allow these animals to breed in the way they wish; from now on, we decide when and with whom they will breed. The purpose of this is to select traits, both physical and behavioral, that we prefer; thus breeders "cull" animals they do not want, especially aggressive animals—breeding the gentler animals with other gentle animals. (Remember, we are not talking here about the later breeding techniques that go back only a few hundred years.)

In the case of the first wolf, he was probably initially confined in some way to make certain he did not leave the village, though we discovered later on that such confinement was unnecessary: The wolf was happy to follow us on our walks. (On the other hand, early humans probably had no idea what a fence was, so the association may have been voluntary right from the beginning. This can only be a "just-so story"—no more than a guess.) This is a scenario that must have been repeated many times over, or at least we assume so. Once a male and female mated and had cubs, we kept only those cubs most easily tamed. This process happened over and over, and eventually the wolf's very appearance began to alter. Nobody is certain how long this process took, but it was at least several thousand years. Eventually, the wolves' heads became smaller. The teeth became smaller too. A dog was emerging from a wolf. Nothing about this is controversial, even if only partially understood. But something that has never been definitively answered is *when* this took place. It was before other animals, but how long before? That question is now being answered in ways that are very controversial indeed, mostly because they cast light on something other than the age of domestication.

IN THE PAST fifteen years or so, an increasing number of animal behavior scientists have shown an interest in a certain aspect of the dog-human relationship, spurred on by a series of significant genetic studies led by a group of scientists in Sweden and at UCLA that have put the beginnings of the relationship further back in time than had previously been believed (anywhere from 15,000 years to 150,000). These researchers began asking an altogether new question: Could humans and dogs, unique among all species, have chosen a kind of mutual evolution (*self-domestication* is another way of putting it) that would benefit both species equally, and in unique ways? Colin Groves, the prolific archaeologist and anthropologist from the Australian National University in Canberra, in a keynote address entitled "The Advantages and Disadvantages of Being Domesticated," summarized this position as follows:

> *The human-dog relationship amounts to a very long-lasting symbiosis. Dogs acted as humans' alarm systems, trackers and hunting aides, garbage disposal facilities, hot water bottles, and children's guardians and playmates. Humans provided dogs with food and security. The relationship was stable over 100,000 years or so, and intensified in the Holocene into mutual domestication. Humans domesticated dogs, and dogs domesticated humans.*

Whether Groves is correct in his time line is a contentious issue. Until about 1997, the consensus was that dog domestication happened between 12,000 and 15,000 years ago, still well before the advent of agriculture (or even horticulture for that matter). But in 1997 some very sophisticated genetic studies were published (sequencing for the first time the complete mitochondrial DNA genome where the lineages between dogs and wolves were clearly distinguishable, as they had not been in the nuclear genes) that suggested that our time with dogs goes back much further, perhaps as long ago as 130,000 years, though the latest studies of mtDNA I refer to above would seem to contradict this. The very early date is much disputed, and there are serious scientists on both sides—those who say it definitely is old, and those who say it is not. Then there are those who believe a compromise is necessary and suggest a figure of 40,000 years. How much difference does it make? The longer the relationship, the stronger the possibility of coevolution, because the relationship would have been in place as humans were still evolving in significant ways, and while wolves were also evolving into a brand new species: the dog.

PROFESSOR ROBERT K. Wayne, a leading geneticist at the University of California at Los Angeles who has been largely responsible for the new research, wonders what characteristics dogs had that caused them to have such high value in ancient societies, and why were they domesticated thousands of years before other animals and plants? Did dogs contribute to the rapid expansion of humans into the New World? Dogs definitely changed early human societies. Most speculation has been about hunting, but I suggest that far more profound was the effect dogs had on our ability to feel love, affection, and friendship.

Wayne believes the evidence suggests a very long coexistence of humans and dogs. Genetic analyses support this conclusion.

He points out that dogs have been living in close association with humans much longer than any other domestic animal or plant species, even if we stick with the 15,000-year date instead of the 100,000 one.

A recent scientific article, "From Wild Wolf to Domestic Dog," sums up our current knowledge of those issues so important to the understanding of the origins of domestication:

> *The dog (*Canis familiaris*) was the first species to be domesticated. This event was a crucial step in the history of humankind and it occurred more than 15,000 years ago when humans were generally nomadic hunter-gatherers. Dogs were domesticated at least several thousand years before any other plant or animal species, and the few ancient remains found so far come from Europe, North America, and the Near East, suggesting they rapidly spread throughout the world after initial domestication events. As a result of the scarce and highly fragmented archaeological evidence, little is known about the specific location, conditions, or causes of domestication. Knowledge of the pattern and process of domestication is essential to understanding human civilization at the end of the Stone Age and the transition from hunter-gatherer to agrarian societies.*

With this quotation as background, I am able to identify at least five theories that are part of the new paradigm that lies at the heart of this book. The first is by David Paxton. A veterinarian from Tasmania, Paxton argued in a PhD thesis that the very early association between dogs and humans perhaps even allowed humans to dispense with smell as an important sensory modality, and develop language instead. The second is by Jon

Franklin, who in his fascinating book *The Wolf in the Parlor* carries Paxton's argument further:

> *What could be more stunning than the idea that modern humans appeared at the same time as the first dogs? [. . .] At precisely this geological moment, some twelve thousand years ago, the human lost nearly 10 percent of its brain mass and, in the process, became the animal destined to build civilizations, pyramids, and spacecraft. And as that new, smaller-brained but somehow smarter animal walked out of the swirling fog of time, it was not alone. It was accompanied by, of all things, a mutant wolf.*

Number three is Meg Olmert, who in *Made for Each Other* stresses the role of oxytocin, so powerful in nursing mothers, and the feel-good chemical serotonin—arguing that being around wolves (dogs) bolstered our oxytocin system. And it works both ways: "Making more affectionate animals makes more affectionate people. This is the biology of bonding in a nutshell. In a very short period of time, a single neurologically inspired human preference for affiliation could have changed the shape and fate of wolf, dog, and humanity." (She describes this as having taken place "sometime between forty thousand and fifteen thousand years ago.")

Number four is the view of the Austrian researcher Wolfgang Schleidt, who holds a very similar position to my own. Finally, Paul Taçon, professor of anthropology and archaeology at Griffith University in Queensland, in an article (written with Colin Pardoe) entitled "Dogs Make Us Human," wonders whether the very idea of "mateship" might not come from the dog/human relationship. He notes that in some happens, it is rare enough to make the news. There is nothing newsworthy, on the other hand, about a dog who falls madly for a pig, cow, elephant, cat, or—most prominently and commonly—one of us.

The exact mechanism of this remarkable mutual evolutionary development in tandem is not known. But no doubt it was facilitated by the fact that both dogs and humans are intensely sociable species: Dogs make friends with other dogs easily, and humans easily form friendships with other humans. So the template was already in place. But one could ask: Why would it so easily generalize to other species beyond the two? Why not keep it exclusive? Was it dogs or humans who first had the idea to extend friendships to other species as well? It is impossible at this distance to say, but I believe that the primary example of dog/human friendship naturally suggested extending it. If we could do it with each other, why not with other species as well? Would wolf pups, introduced to a small animal from another species, play gently with the animal? It is unlikely. They are carnivores, after all. "So are dogs," you might object. Yes, but they are carnivores who have been encouraged by us to play gently with the young of our species right from the beginning, even if not always successfully.

Why do I call this mutual evolution? Because it is not seen anywhere else, and the most parsimonious explanation is that we encouraged and facilitated it with one another. This is, I admit, only a hypothesis, but it is one that fits the facts remarkably well, and until something more convincing comes along, I believe it helps us to understand what is otherwise an anomaly in nature. Humans were not completely human 100,000 years ago. We did not have language, for example. So the disparity between humans and wolves would have been even less than it was

northern Australian Dreamtime stories, it is the dingo that makes us human. He ends his article by noting, "The human-dog partnership set both on a course that would change the world forever. Indeed, it can even be argued that the success of our first relationship with another species laid the ground for the eventual domestication of other animals, ultimately leading to a radical transformation of the planet."

My own thesis builds on all these views and carries them further, to claim that dogs and humans together developed a far greater capacity for interspecies love (with one another and with other animals as well) than any other two animals on earth. All six of these views share many of the same positions. Soon it will be time for the first conference on wolf/human coevolution.

My life with Benjy has been like a three-dimensional model of this very thesis. Many of you too will have experienced something very similar living with dogs of your own. There is nothing very startling in making these claims, and yet it has set up a new paradigm for the understanding of both dogs and humans, and of the coevolutionary theory that brings in a new recognition of the oldest of all human-nonhuman bonds. The theory and its practice revolve around love—the love of dogs for us and our love for them. It is worth considering in greater detail the nature of this love when it comes to dogs, for there are those (dogless) people who still doubt it should carry such an exalted name.

SO THEN, HOW can we characterize this love and what exactly is this quality that Benjy seems to possess in abundance, that he has absolutely no problem showing? Nobody who sees or experiences this love could for a moment doubt its genuineness. As people who live with dogs know, this is not unique to Benjy. Nor is it some mysterious property of golden Labs (though it is the rare Lab who does not show such a quality). It is not surprising that given this gift, dogs and humans should come together to display it to each other.

Most people do not need to be convinced that humans have a very special relationship to dogs. Just consider: Have you ever seen a dog in a zoo? Of course not. There are no dogs in zoos (except dingoes). That is because we have no relationship, at least none that is personal, with animals in zoos, whereas our relationships with dogs are intensely personal, the most personal relationships we are ever likely to have with another animal. We look at our dog and our dog looks back at us. In a zoo, animals avoid our gaze. They don't even see us; in fact, they may not even notice us. We are simply wallpaper for them, not even particularly annoying wallpaper. It is a complete perversion of the ideal relationship: dog and human.

DOGS HELPED MAKE us human, which they did by reciprocal love, teaching us about love and empathy, especially love and empathy for a member of a different species. It is not a mere coincidence that only two species on our planet have a remarkably similar and consistent ability to form friendships across the species barrier: dogs and humans. I believe we developed this ability in tandem, that we evolved together in one another's presence. Humans form friendships with dogs, but also with many other animals, and dogs do exactly the same thing. While it is unusual for any domesticated animal except the dog to form strong bonds with an animal of another species, it is

common for dogs to do so. Some people might contend that they are simply imitating us. Why not, others could argue, see it the other way around: We are imitating dogs. Both are right and wrong. Surely it is a more simple argument: Dogs and humans developed this ability to influence one another simultaneously. Dogs watched us; we watched dogs. In the beginning, we only had one another to love—or rather, members of our own species, then one another. But eventually, after our own bonds were cemented in place and as humans domesticated other animals, we were able to form bonds of affection with other domesticated animals—cats, for example. Sometimes we wanted our dogs to love other animals as well, even cats. We reward dogs for imitating us in this regard. We praised dogs for showing affection, concern, and gentleness for other creatures, and dogs bond even more closely with us as a result of our praise. Everyone profited, including the other species with whom both our species made friends. Surely cats who are friendly with dogs and humans are better off. This is a process that has been going on for some thousands of years, which is why it has been so successful. In theory, we could probably achieve the same results with pigs or other domesticated animals such as goats and possibly even cows (though people seem to have difficulty in bonding with cows). But we have been associated with these animals for too short a period to make such a profound change. And, even more important, for various reasons we have not been as willing to open ourselves up to these other animals in the same way we did with dogs. That may be changing now, but for most of our history associating with domesticated animals, only dogs and cats became intimate companions. Pigs will certainly tolerate other animals (most wild animals will as well), but when it comes to forming close bonds, if and when that

to become, and the communication between humans and wolves could have been more like a communion than order-giving, as it so often is today.

Affection, bonds of friendship, even love were present from the very beginning. The earliest surviving archaeological evidence of domestication shows quite clearly the skeleton of a human (whether man or woman is not clear) holding a small puppy or wolf cub close to the breast, clearly meant to demonstrate the close association of the two in this world and hopefully in the next.

This mutual domestication is unique in nature. In short, dogs and humans domesticated each other when it came to affection, perhaps even love, and certainly enjoyment and tolerance of alien species. When I told my theory to the geneticist Anatoly Ruvinsky (who was part of the original "fox farm experiment" in Siberia and is now professor of genetics at the University of New England in Australia), he told me that it was perhaps a rather poetic way of stating it, but I was not wrong. The relationship between the two species is unparalleled in nature, he believes. In fact, he would go so far as to say, and I quote him, "Humans would not be the humans we are without dogs."

We know that for at least the last 15,000 years and continuing down until today there have been hardly any human habitations without dogs. Whether it was a single event, or one repeated in different places, the fact is that wolves were turned into dogs from a very early time, and just about everywhere there were humans. For example, the only domestic animal of the Janwas of middle Andaman Island, still hunter-gatherers today, is the dog. And tribes without a single other domestic animal, such as those in Central and South America (llamas, turkeys, and the guinea pig were to come later), nonetheless have dogs living with them. Even the skeptic Stephen Budiansky, at the beginning of his book *The Truth About Dogs*, writes: "Before there were cities or even villages, before there were farms, before there was writing, before people could afford the meanest luxury, before people fretted about stress, before humans were indeed scarcely human, dogs latched on to human society, survived and flourished." Over thousands of years humans and dogs coevolved, learning about trust, devotion, empathy, and other positive emotions from one another. If we are capable of having these same attitudes toward other species (curiosity, a desire to be near them, to nurture them)—even those we have long exploited—then so are dogs. Not everywhere, not always, not universally, but often enough, and in enough different places that today we see almost a new species, or rather two ancient species that grew an extraordinary new capacity in common: the desire to live with other species in a new kind of harmony, absent exploitation.

Wolfgang Schleidt, chair of animal ethology at the University of Vienna, who holds the closest approximation to my views of any other scholar, has written an article with Michael D. Shalter that describes the early relationship between humans and dogs. They write, "The closest approximation to human morality we can find in nature is that of the gray wolf, *Canis lupus*." They point out that wolves provision not only their own young, but also other pack members; they babysit and do other things found in human societies, including—most important—protecting the entire pack even at the expense of their own lives. Schleidt and Shalter propose that initial contacts between wolves and humans were mutual and subsequent changes were a process of coevolution: "The impact of wolves' ethics on our own may well equal, or even exceed, that of our effect on wolves' changes in becoming dogs." They claim that there is something in the bond between humans and dogs that goes beyond that between us and our closest primate relatives, the chimpanzees (italics in original): "Here *we are not talking about intelligence*, but about what we may poetically associate with *kindness of heart.*" They point out that we not only value a dog's intelligence, but his or her warmth, affection, playfulness, and loyalty. Did we humans then learn traits inherent in wolf society from dogs?

WHAT ARE THOSE traits? First and foremost would have been the very ability to attach ourselves to a member of another species. This had never happened in the history of humans or in the history of wolves. Wolves played only with wolf companions; wolves cohabited only with wolves; wolves hunted only with fellow wolves; wolves communicated only with wolves. (We know this from modern studies of wolves: Wild wolves still do not form bonds with other species.) Humans too: We had no history of any kind of association with another species that was not predatory in nature. We have *myths* of such associations, but there is no evidence at all that humans ever bonded with any wild animal other than the wolf. We were either predator or prey, the hunter or the hunted. We had no friends in nature beyond other human beings. And then suddenly, this changed. Human children *made friends* with a member of another species: a wolf cub, in the scene we imagined earlier. Imagine how extraordinary this must have seemed to the adults looking on. For children it may have felt natural, even if they had never done it before, but to the adults watching, it was something unheard of, something unique, something they were witnessing for the first time.

But it obviously appealed to both parties, humans and wolves, because the practice flourished. And from this initial companionship, this mutual attachment, would come other qualities to be found in both species—perhaps uniquely in both species. Humans began to feel for their companion wolves, and vice versa. They began to anticipate the presence of the other and to take pleasure in being together. It still astonishes me that if Benjy is lying in our bed, I always know when Ilan is coming down from his bedroom into ours, because before I can even hear any steps, Benjy's tail begins to thump. He does not budge. There is no movement anywhere in his body except for the increasing frenzy of the tail thumping. Benjy hears Ilan before we do, and the pleasure he feels is communicated to his tail and from his tail directly to us. We know he is happy.

DID IT HAPPEN like this with wolves? Wolves don't normally wag their tails, except low down and in greeting. But at some point it must have become clear to wolves that they could most easily communicate their pleasure to us by so doing, and that we would have no difficulty in interpreting this gesture as one of happiness. I don't think it was ever a conscious decision, and it must have been very gradual, taking perhaps hundreds or even thousands of years. But soon humans and wolves understood one another perfectly.

Just as wolves could experience happiness in our presence and know that we felt the same, they must also have begun to experience our pain as well. At some point the first wolf empathized with the sorrow of his human companion. Perhaps a human weeping somehow communicated to the wolf the sorrow involved and the wolf reciprocated with an attempt to alleviate the suffering. Licking, groaning in empathy, some sign that they understood and felt bad. Well, this was new! No human had ever had the experience of an *animal* showing sympathy. The first time must have seemed like a miracle or some sort of supernatural occurrence. But it caught on and spread like wildfire. We stroked our wolves when they were in pain and they licked our hands when we suffered. We reinforced the behavior in one another again and again and again, over many thousands of years, until it was instilled in both species: Thou shalt sympathize with one another. Of course when I speak of "wolves" I really mean the wolf that was turning, or had already turned, into a dog. The wolf, even a tame wolf, is still a wild animal, far removed from a domesticated dog. This has been shown conclusively by the research of Vilmos Csányi at Eötvös Loránd University in Hungary. Together with his graduate students, he was able to show just how different dogs are from wolves, with respect to humans: At five weeks of age, wolf cubs were introduced to a room containing their hand-raiser and an adult dog, both sitting motionless, and a human staring into space. Mr. Miklósi shows a video of what happened: A gawky wolf cub stumbles awkwardly up to the dog, sniffs it a bit, then does the same to the human before climbing into the person's lap and going to sleep. No eye contact is made with its caregiver; the cub appears to treat the person like a comfortable piece of furniture. Mr. Miklósi's next video shows a dog puppy wandering into the same situation. It too wanders over to the dog for a sniff, but then waddles over to its caregiver, stares it in the face, and begins yipping for attention. When the caregiver remains motionless, the dog wags its tail, barks, and begins licking the person, trying to establish contact. It then sits down in front of the caregiver, ears up, apparently waiting for contact. This is what thousands of years of interaction at a deep emotional level has achieved. It is, by any standards, remarkable, and possibly unique.

How did we achieve this emotional closeness? One way was by including the wolf, that is, the soon-to-be dog, into the human family. We name members of our family, and dogs are no exception. Would ancient hunter-gatherers have given names to their wolves? Of course. Would wolves (as they became dogs) have come to know their names? Of course. (Wolves today, however, do not.) Benjy, like most dogs, knows the names of his favorite companions. So when I tell him, "Run up the path, Manu is coming," he races forward, looking for his friend Manu. He knows that Manu stands for the little eight-year-old boy he plays with, and is now searching for him with eagerness. There is no reason to believe this was ever different. Giving a name to another person was common; giving a name to another species was yet another first. It had never been done before, but once it happened, it seemed obvious and was quickly imitated. Soon, every domesticated wolf (that is, every dog) would have a name: his or her very own name that implied a special relationship with the person who bestowed it. (Dogs may not have names for us, but they nonetheless know our names.) Research by Michael Tomasello, co-director of the Max Planck Institute for Evolutionary Anthropology in Leipzig, has demonstrated that dogs are far better at reading social clues from humans than the great apes, even though apes are much closer to us genetically. Tomasello explains that chimpanzees, our closest relatives, can follow a human's gaze, but they do very poorly in a classic experiment that requires them to extract clues by watching a person. He suggests they are hardwired for competition, whereas dogs are hardwired for cooperation, especially with us.

ALL WILD ANIMALS sleep only with their own kind. This was true for humans as well until fairly recently in our evolutionary history. But now here was somebody else with whom one could share the bed: a wolf. They were warm, reliable, and cuddly. They were both a night blanket and a night protector. Humans felt much safer sleeping with dogs than they did sleeping alone. We still do. When Benjy is by my side, there is no need to lock the door. In the beginning, like many families, I had rules: Benjy was *not* to sleep in our bed or in the beds of our boys. The rule did not last long. I am not even sure why I had it in the first place. Perhaps I had read somewhere that a dog who sleeps in your bed usurps your rightful place as head of the pack. But as I said earlier in this book and in many other books, I am not a great fan of hierarchy between humans and other species. I don't believe in rightful places and heads of packs. I like equality. Benjy does not take advantage of my easygoing nature, and there is no need to be strict with him or impose rules that deny both of us the pleasure we get from each other's company. So Benjy sleeps in whatever bed he feels like.

This is a form of trust, and it must be accepted on the part of both wolf and human. For all animals, the most vulnerable self is the sleeping self. For a wolf to sleep with a human is to indicate his complete trust. He has to believe he is safe. And for the human, a natural fear must also be overcome: We can only imagine the mother whose child first sleeps with a wolf. The anxiety must have been enormous. "Are you crazy, letting her sleep with a *wolf*?" But sleep they did, and soon it must have become a fad—everyone had to sleep with his or her wolf! Of course, on cold winter nights, dogs would have kept us warm. They are still used in this way by Aborigines in Australia. Some sleep researchers have noted that dogs and humans who sleep together have similar biorhythms. This might explain why some dogs resemble their "owners." This mutual biorhythm is a promising new field of research in sleep physiology, as the Sleep and Behavior Medicine Institute noted in 2009.

Whether we are humans or wolves, we want to protect our newborns from any harm. At some point, the first wolf gave birth in the shared shelter of her human friends, and then allowed them to play with her cubs. Equally, at some point, the first human gave birth to a newborn and allowed the family wolf to sniff and lick the baby (or am I being naïve here?). Other animals do not give birth in our presence, and we do not trust other animals to play with our newborn children. Perhaps we would not go so far as to leave our babies in the care of our dog, but many societies did. There is an entire genre of literature around this theme: Parents leave their sleeping child in the care of their beloved dog. When they return late in the afternoon, the dog rushes out to greet them, and he has blood on his mouth. The father, without thinking, is convinced the dog has eaten the baby, and in his blind rage kills the dog. When they reach the house, the baby is lying peacefully asleep and next to her crib is the dead body of a large poisonous snake.

Here is the most famous of these stories, from Wales, in the early thirteenth century: One fateful day, Gelert refused to accompany Llewellyn, his master, but instead ran howling back to the lodge. When Llewellyn returned, his dog went bounding up to meet him. His muzzle was splashed with blood. Llewellyn entered his sleeping quarters and found a scene of confusion with rooms disordered and articles scattered in heaps. Llewellyn had a son, barely a year old, and as he recalled how Gelert and his little boy used to play together, a terrible thought came to his mind. He rushed to the nursery only to find the cradle overturned and the bedclothes bloody. Though he looked frantically for his son, the child could not be found.

Turning to Gelert, whose muzzle was still wet with blood, Llewellyn went into a great rage and cried, "Thou hast killed my only son." He drew his sword and drove it into the heart of the hound. Then—as all was silent but for the steady drip of blood onto the stone flag floor—the wail of a baby could be heard. On searching further, Llewellyn found his son safe and well, lying next to the body of a large gray wolf.

I would like to think this is only a myth and has never happened, but stories like it are so consistently reported across cultures that I believe it may have happened many times over the years and in many different cultures. Perhaps, though, it is simply a universal indication of the great trust we place in our dogs and how very much deserved and earned is this trust.

DOGS CAN ALSO share our fears. Perhaps too they are the only other domesticated animal who senses when we are not there and wants us to be there. They are just as worried about our getting lost as we are about them getting lost. A few days ago, Benjy went bounding up the steep path leading to our street because he thought the rest of the family was returning. I thought so too. When I realized they were not, I turned around and started back down the hill. Benjy refused to walk with me, so I let him wait in the middle of the path since I assumed they were coming any minute. But they were delayed. When Benjy did not come home after an hour, I walked up the path to find him, but he was gone. The horrible feeling came into my throat that anyone whose beloved animal is suddenly missing will recognize. Our heads fill with terrible thoughts of all that could happen to our animals. I am convinced that Benjy has the same thoughts about me and my family when we are, in his terms, missing.

How do I know this? Well, think about abandonment. Every human is frightened of it and I am convinced that every dog is as well—especially if he has had any experience of it. Benjy did not know the reasons he had to move from family to family, but he understood that something was wrong—that somehow he was found wanting. Or maybe even that concept is too big a stretch. But whatever his conception of it was, his recognition *at a feeling level* that he had been abandoned was correct. I am sure most of you have had the experience of walking through a dog shelter. There are dogs in there, especially older dogs, who know that their fate is sealed. They have that look in their eyes. They have given up hope. They are aware that they have been abandoned to their fate, and it is not a good one. They sense the worst. Nobody who works in a shelter has ever denied this to me. They know that the dogs know that they will die alone and unloved. We often wonder whether other animals have a concept of death. I think they do, and certainly research about elephants indicates that these animals are acutely aware of death. But we need not wander so far afield to answer the question. Dogs in shelters give every indication that they understand what is about to happen to them, which is one of the reasons they make themselves as adorable and appealing as possible to anyone who enters. They are literally fighting for their lives.

ONE OF THE reasons that we attribute feelings to animals in general is because of our relationship with dogs. It is not, strictly speaking, anthropomorphic to speak of dogs feeling very similar emotions to humans, because we observe this happening daily in our relations with dogs. We are not projecting; we are simply observing, and have been doing so for thousands of years. Steven Mithen, professor of early prehistory and archaeology at the University of Reading, with an interest in psychology, has persuasively argued in a series of books that "the first modern humans, and especially those after 50,000 years ago . . . developed that diverse range of relationships with animals that we see in the modern world today. One of the key mental developments lying behind these new relationships seems to be that of anthropomorphizing animals." I agree with his argument, and would only add that this ability to enter into the mind of other animals is undoubtedly due to our intimate relationship with *one* animal, the dog. Dogs allowed us insight into the animal mind because they allowed us insight into their minds.

RECENTLY, SCIENTISTS HAVE been intrigued by a border collie, Betsy, who has a vocabulary of more than 300 words. "Even our closest relatives, the great apes, can't do what Betsy can do—hear a word only once or twice and know that the acoustic pattern stands for something," said Juliane Kaminski, a cognitive psychologist who is now studying Betsy. Kaminski and her colleague, Sebastian Tempelmann, went to Betsy's home in Vienna to give her a fresh battery of tests. "Dogs' understanding of human forms of communication is something new that has evolved," Kaminski said, "something that's developed in them because of their long association with humans."

Are we giving undue importance to language as a tool of communication simply because *we* employ it? After all, there are many children born deaf and blind who communicate at a sophisticated level without words. Dogs too communicate with us (and with each other) at a very advanced level without the use of words. We share unique traits with dogs now perhaps because of our prolonged contact, but also possibly because we are so similar to dogs in our social lives. No other animal is as similar to us in a number of important ways.

Only dogs study our faces and make prolonged eye contact with us. They puzzle over our intent (dogs have that special quizzical look indicating their attempts to understand us). Staring at another animal is almost always considered a sign of hostile intent. (Always glance away from a cat if you wish to reassure her you mean no harm.) We can also stare at dogs without their misinterpreting it: We gaze steadily with affection directly into the dog's eyes. We can even immobilize their heads while we do this (don't try this with your cat). They correctly read us. Dogs want to read us. They are constantly attempting to understand what we feel from how we look and our body language. Miraculously, they can often read us better than we can read ourselves. Hardly any other animal seems to care how we feel, and certainly no other animal is as alert to our emotions as dogs. Wild animals rarely have a clue: When I tried to approach a herd of elephants, attempting as best I could to indicate "I come in peace," I was charged and nearly killed by the matriarch. She failed or did not bother to read my emotions or intentions!

I HAVE WATCHED a dog fix on his "master's" face, body language, gestures, and more, to follow his instructions to the letter. My friend John Friedberg, a neurologist and author of the single best book ever written against electroconvulsive therapy (ECT), *Shock Treatment Is Not Good for Your Brain* (one of the most profound reading experiences of my life), lives with Buddy, a Yorkshire terrier. Buddy is one of the most attentive dogs I have ever met. Their relationship is not dissimilar to a good marriage—that is, the perfect meeting of two fine minds, with a visibly deep bond. They spend every day and night together and what they each feel for the other is surely love. If the word has any meaning—then this is love.

We went for a walk on the Berkeley, California, marina. What struck me right away was Buddy's acute awareness of every word John spoke to him. When he threw the ball, Buddy would race after it, and John would direct him: "Right, left, keep going, straight ahead—no, behind you!" Buddy stared at John's face and tried, almost always successfully, to make sense of any words directed at him (he ignores the ones between me and John as gibberish). Watching Buddy, what strikes me, as the core and the source of this ability he so strikingly possesses, is the relationship between these two intimate friends. Buddy *wants* to understand and thus has acquired the ability to do so. Of course, this relationship that I so admire may be necessary, but not sufficient, to produce such astonishing results. Benjy has a similar attachment to me, but is perplexed by the simplest verbal commands. Mind you, he is really just not that into obedience, so it could be that his cluelessness is most convenient. He has trained me, as one dog scientist told me, to not annoy him with pointless commands. How did this ability that Buddy and other dogs have evolve? Only one way: by association with us. Wolves, we know, will not search the human face for clues, even the tamest wolves. Of course, it is a two-way street. Some people would make fun of a highly accomplished neurologist who is close friends with and has a deep commitment to a 7-pound dog. Not me. John has a profound relationship with Buddy that gives both of them intense pleasure as well as intellectual rewards. John knows the inside of the mind of a different species, something most people will never experience. They are companions in every sense of the word. Buddy, *mi compañero*—what word could be more appropriate? What other two species on this planet can engage in a friendship of this intensity? My thirty-five-year-old daughter, Simone, a veterinary nurse, tells me she has this with her nine-year-old cat, Mr. Jazz. He follows her everywhere and prefers her company to that of any other, human or animal. But he is unreliable and might well attack her arm, should he feel so inclined. It is as if cats, no matter how attached to us they become, are forever fighting an internal battle: love, says their experience; detach, says their nature. What I saw in the walk with John convinces me yet again that among the most interesting mysteries on our planet is the seemingly everyday occurrence of one of the deepest bonds that exist between any two people. It is, in part, unfathomable, and in equal part the very core of our nature and the nature of dogs.

But why should dogs be so savvy about what we feel? One reason is that dogs have had thousands of years to recognize that when we look into their eyes we are not about to attack them. They have come to understand the gesture as one of affection, even if it is not such a gesture in their own world (a dog won't stare at a strange dog, but will often make more gentle eye contact). Surely one of the great mysteries of the animal world is why dogs and only dogs will permit this degree of intimacy, one that goes against their very nature. The only explanation I can think of is that they love us. Jacqueline Sheehan ends her novel *Lost & Found* with this lovely passage: "But in this life he is dog. His life is ocean, stick, ball, sand, grass, ride in the truck, sleep by the bed, look deep into the eyes of humans, lure them outdoors, greet them with a burst of joy when they come home, love them. Fill this brief life with more."

DOGS APOLOGIZE, JUST as humans do, when they unintentionally hurt a canine companion during play sessions. This was discoved by Marc Bekoff, who noticed that when a dog makes a play bow and begins to play, he sometimes accidentally hurts the other dog. The dog who causes the injury then play bows again, which is correctly interpreted by the second dog as a form of apology, almost always accepted, and then play is renewed. Bekoff notes that play bows are often used as a signal to say "let's continue playing," especially when the play mood might otherwise end. (Gisela Kaplan, professor of animal behavior in Australia, told me that she has observed this apology behavior closely, and has found that dogs apologize *within seconds*. Notoriously, dogs do not bear grudges and this may be the explanation: They are prepared to immediately apologize if they make a mistake.) In a recent article, Bekoff writes, "A dog or wolf may cock her head from side-to-side and squint, as if she is wondering what went 'wrong' when a play-mate becomes too assertive or too aggressive. Perhaps they feel indignant when they are wronged, when their expectations or sense of justice is violated, when they feel they are not being treated 'right.'"

DOGS ARE NOT the only species to apologize. Some birds do and some primates do as well, as noted in the endnote above. And of course, humans—at least some—apologize over and over. They have to. Other canid species apologize, including wolves and coyotes, so it is not likely that dogs learned to do so from observing humans. But did we learn from dogs? Is it possible that humans had some inkling of the ability of dogs to play without harming each other, and learned the value of an apology? Certainly dogs understand our apology. They don't look for a play bow, of course. They look at our eyes. If I accidentally step on Benjy's tail, he gives out a yelp, and I immediately tell him over and over how sorry I am. He checks out my expression and my voice to see if it was intended or not. Then he licks my hand to let me know that he understands that I did not step on his tail on purpose.

Dogs are notoriously sensitive in this regard. Perhaps they have had to be: There are and always have been humans who do intend them harm. Dogs have a terror of being hurt. When they think you are after them to harm them, or even when you simply step on their paw or tail by mistake, they look absolutely stricken and terrified. It is like a primal fear. It *is* the primal fear. Indeed, how could it be otherwise? All sentient beings (a class to which we belong, obviously) fear their physical disintegration above all things. Ernest Becker in his classic study, *The Denial of Death*, was right, I believe: There can be no greater fear than the fear of death. Any harm to the body, any threat to our physical integrity, is the first step toward nonexistence. *All* animals run from the slightest indication of approaching harm. That is what flight distance is: a calculated sense of how safe you are. Dogs have managed to narrow the time when they doubt us to practically zero, and of course they have *no* flight distance from us at all. Apology is the opposite of the denial of death; it is the affirmation of life, and I find it fascinating that both humans and dogs indulge themselves in apologies on a constant and permanent basis.

Cats, by contrast, rarely apologize. If they happen to hurt another cat, the encounter usually ends in a fight, not an apology. They don't have the same signals that dogs do because, as a solitary species, they never needed them. Why, after thousands of years with humans, cats have not learned to play with us on our terms is not clear. Perhaps they feel it is enough that they grace us with their presence! As far as I know dogs do not apologize with a play bow to humans they accidentally injure. They lick, and they look abashed.

It is always risky to write a sentence that begins, "No species in the wild . . ."—in this case finishing with "plays with a member of another species." (Of course, species in the wild play with one another all the time—more frequently when they are young, but even when they are mature—and some species such as dolphins seem to play all the time.) I say it is risky because we simply don't have sufficient knowledge of all animal behavior in the wild. That said, I have not encountered in my reading reliable observations of animals playing with a member of another species. Juvenile animals could well do so in theory, but in practice are rarely given the opportunity. It is impossible to answer the question of whether they have any interest in doing so, should the opportunity arise. Play, even with members of the same species, is not without its risks. Any injury could well spell death.

Obviously, over the thousands of years we have been together, dogs have attempted to communicate more effectively with us, just as we have with them. One way is by barking. We may well be slouches when it comes to deciphering this form of communication.

Only dogs bark: No canine living in the wild does so as often as dogs. Occasionally wolves and coyotes will bark in the wild, but the typical dog bark (whether deep, baying, or yapping) is peculiar to domesticated dogs. Coyote pups and wolf cubs do it as well. Another neotenic trait! The purpose of barking in dogs is not at all clear to me. We hear only one sound, but it is possible that there are many variations in barking with different meanings, and even an attempt to communicate. Geneticist Anatoly Ruvinsky suggested to me that barking is rather a consequence of domestication than a "purpose-developed" trait and better seen as a by-product. Still, if wolf *and* dog howls reveal distinct individuals (like a signature whistle in dolphins), the same must be true for barks. Certainly humans claim to recognize their dogs' barks—I can certainly tell it is Benjy barking rather than a neighbor dog. I would be surprised if we don't turn up a communicative function for barking in the near future.

IN THE LAST five years there has been great interest in the question of mirror neurons, cells in the human brain that fire when we observe people undergoing experiences with which we can identify. Watching somebody choke up and begin to weep in sorrow triggers mirror neurons in our brains that simulate (or reflect, as in a mirror) the person's distress, and we feel empathy because we are literally undergoing a version of the same experience. The prominent neuroscientist V. S. Ramachandran famously prophesied that "mirror neurons will do for psychology what DNA did for biology." In reporting on the new field of research, the science journalist Sandra Blakeslee notes that some experts believe that monkeys, apes, and possibly elephants, dolphins, and dogs also have rudimentary mirror neurons. Recently, a video of a dog rushing into traffic to drag a severely injured dog out of harm's way was viewed hundreds of thousands of times. There is no evidence that the dogs knew each other, so surely the instinct to rescue required mirror neurons.

But if chimpanzees have mirror neurons (nobody is yet certain), the question arises of why did we not form relationships with chimps. Is it because dogs use their mirror neurons not just for other dogs, but also for us? If so, how did they develop this special capacity? After all, many humans lack the ability to feel empathy for the suffering of others (remember the famous maxim of François de la Rochefoucauld: "Man has an infinite capacity to endure the suffering of others"). Could it be that dogs in fact have a more advanced version of mirror neurons than do humans?

At some point, this empathy and sympathy, simple companionship and the pleasure in one another's company, sleeping near to each other and knowing each other's names, passed over into something much deeper. I will call it love. It was probably called that right from the beginning. At some point in our history, no doubt a child told his father, "But Father, I love my wolf. I do not want to be parted from her." And the word would have been familiar to humans from its use for other humans. The word for *love* is old, and probably goes back to the origins of language. Men and women used it; parents and children used it. Why not expand its use? Why not include wolves? Well, we did, and the result is clear today: There is no greater love between humans and any other species than that between humans and their dogs. And it works the other way too, just as it did in prehistory: Dogs love humans as much or more than they love any other dog.

Dogs made us human in key ways. They helped to teach us, or at the very least reinforced, how to love, to empathize, to trust, to play (as adults, not just as children), to apologize—even, in certain ways, how to communicate. In the chapters that follow, we'll explore each of these, along with a devil's advocate consideration of what happens when dog-human interaction turns hostile.

CHAPTER 4

ARE DOGS UNIQUE? COMPARING DOGS TO OTHER DOMESTICATED ANIMALS

Throughout this book, I have compared dogs to other domesticated animals. The reason I do this is because I am trying to make an important point: While we have domesticated many other animals, dogs are in a category by themselves. They are not simply one among many domesticated animals. It is for this reason that we have allowed ourselves to be transformed as a species by dogs, and only by dogs. No other domestic animal has influenced us in this profound manner.

It is worth spending a little time on the history of domestication in general. When we think of domestication, we often separate out the domestication of the wolf into the dog for the simple reason that it took place so long before any other domestication so we almost consider it a distinct incidence. In all other cases of domestication, we are referring to the time humans first started living in permanent or semipermanent dwellings. Domestication really begins with early human architecture, about 13,000 years ago. The beginning of agriculture—that is, of the domestication of wild plants—took place about

10,000 or 11,000 years ago. Whatever the reason may have been (climate change—arid conditions were becoming more common, and humans needed or thought they needed a constant supply of meat), animals began to be domesticated only with the advent of agriculture. No animal except the dog was domesticated before about 8,500 years ago. By the seventh millennium BC, the domestic goat and sheep had become the principal source of meat for farmers in western Asia, the so-called "fertile crescent." Cattle and pigs were also undergoing the process of domestication at this time (for this is something that takes place gradually and not in one sudden leap).

IN 1865, CHARLES Darwin's cousin Francis Galton presented a famous paper, "The First Steps Towards the Domestication of Animals," in which he enunciated for the first time the six preconditions animals must have in order to domesticate them: "they should be hardy"; "have an inborn liking for man"; be "comfort loving" (meaning they must not be adapted for instant flight, but permit constraint in a pen); "found useful to the savages" (that is, they must serve as an itinerant larder); "breed freely"; and finally, "they should be easy to tend." It is for these reasons that Galton explains the fact that so few animals have been domesticated, and essentially no new animals have been domesticated in the last 10,000 years, because so few wild animals are preadapted for domestication. Finally, Galton says it is essential that humans learn to communicate with animals:

> *A man irritates a dog by an ordinary laugh, he frightens him by an angry look, or he calms him by a kindly bearing; but he has less spontaneous hold over an ox or a sheep. He must*

> *study their ways and tutor his behaviour before he can either understand the feelings of those animals or make his own intelligible to them. He has no natural power at all over many other creatures. Who for instance, ever succeeded in frowning away a mosquito, or in pacifying an angry wasp by a smile.*

The problem here is that in the case of *all* domesticated animals, the motivation has been one of exploitation. Galton wants us to understand the feelings of domesticated animals to better make use of them; they are there because they serve a human purpose. This is especially true in the case of sheep, and goats, and pigs, and chickens—in fact, it is true in every case of domestication with only two exceptions. It is possible that the two exceptions, dogs and cats, began in the same way as the other animals and were only brought into human society for the various ways they could be helpful or even essential to humans. Cats, once grain came to be stored, were rodent-exterminating machines. Cats could perform this function far better than could any human. Similarly, no human could perform guard duty as reliably as a dog. As sentries around villages, dogs could not be surpassed. My question, and one for which I have as yet found no answer, is this: Why, of all these animals, did only two pass beyond their original functions to become full-fledged family members? In the case of dogs, it is mutual: We are as much accepted into their world as they are into ours. Cats are somewhat less generous, but compared to any of the other animals mentioned (as well as camels, ducks, rabbits, ferrets, geese, reindeer, silkworms, chickens, turkeys, llamas, alpacas, guinea pigs, donkeys, yaks, water buffalos), they allow us greater entry into their private worlds.

The other animals, *all* other domesticated animals, resist us in ways that preclude us from ever becoming intimate with them in the way we have with dogs and cats, except rarely and in exceptional cases (I was close to a duck when I was young, and many people, especially recently, have become very attached to chickens and vice versa). I am intrigued because, according to Galton, we should long ago have learned to exploit precisely those six characteristics that preadapt animals for domestication to the point where each and every domesticated animal is an intimate part of our society *apart* from the gain we can have from them. Just for their own sake. I suspect animals resist this intimacy because they sense our lack of sincerity. I cannot blame them. No less than the world's leading authority on domestication, Juliet Clutton-Brock, ends her book with a terrifying admonition: We must become *more* exploitative, not less, to prevent extinctions: "If, instead of encroaching on the last reserves of the natural environment in efforts to provide more farmland, the indigenous faunas and floras are exploited. Additional animals will have to be domesticated and the elephant, the zebra, and the eland must join the ox and the sheep as providers of meat, leather and other necessities for the welfare of the master predator." These words, I must confess, give me the chills. The proud and mighty elephant has never been domesticated, only captured or forced into slavery. The same is true of the freedom-loving zebra and the free-spirited antelope. They do not want to be turned into leather couches or meat. Nor, would I argue, are these necessities at all; on the contrary, meat and animal products have helped to bring us to the current dead end we now face with global warming, as I (along with many others, including Al Gore) have tried to show in my recent book, *The Face on Your Plate.* We can no longer afford to be *any* kind of predator, and least of all the *master* predator. Dogs, I maintain, have taught us the value of entering a new and different kind of relationship with our evolutionary cousins with whom we share our fragile earth.

I am hardly alone in this desire to see a new paradigm. The Internet is flooded with harmonious images of wild animals behaving in ways that Galton could not anticipate, whether playing with a member of a different species, saving a member of a different species, or just quietly hanging out with a member of a different species. It may be that such activities are only now visible, but there is a different quality to the human desire for such images. We want them so badly we are willing to invent them (as in the charming humanoid dragons of *Avatar*), but we also look for them more intensely and with greater focus. We long for role models in the natural world that goes beyond predator/prey, even as we decisively cast off the yoke of master predator. Our concern now is to understand and bond rather than conquer and kill. I believe we have achieved this paradigm shift because of our proximity to dogs.

We have come closer to dogs, and they have come closer to us, than any other two species have ever managed in history. I would go so far as to suggest that we are no longer two entirely separate species. There is a sense in which we have merged. When dogs give birth in our homes, they treat us as doulas, midwives, intimate family. They allow us to handle their newborns at will. They want us to see them and touch them. They want us to protect them and help take care of them. They enjoy our pleasure in them.

DOGS ARE THE only domesticated animals who cannot survive on their own. In spite of appearances, research shows that dogs are unable to revert to their wild state with even moderate success. They are so wed to us that in our absence they simply cannot thrive. Every other domesticated animal can—and does—go feral, more often than not successfully. Alan Beck in the second edition of his classic *The Ecology of Stray Dogs*, states:

> *My own observations since my original study found that one adaptation of unowned (stray) dogs in an urban environment is to behave like socialized pet dogs. In that way, they are indistinguishable from owned straying dogs and are tolerated as loose pets and not wild dogs—a form of "cultural camouflage." Loose or straying pets and stray (feral) dogs are different. True stray dogs form somewhat stable packs, usually between two and five dogs, and are more active at night and cautious about people. In general, straying pets have smaller home ranges and are active when people are. They usually associate with fewer other dogs—the human family is their significant social setting, or pack.*

Wild dogs, or true stray or feral dogs, are more like wild wolves. I must stress the differences between dogs and "tame" wild wolves. A tame wild wolf is identical to any other tame wild animal: The tame wolf can be affectionate, even loving, fun, and interesting, but the relationship can never be entirely reliable and consistent. At any moment the tamest, most docile wolf (or elephant, lion, tiger) can turn against its human companion and inflict grievous bodily harm. A tame wolf is still a wild animal and completely unreliable (which makes it difficult to understand how early humans allowed them around newborn babies, if they did). The best example of this I have read comes from the visit that Raymond Coppinger made to Wolf Park in Indiana, where Erich Klinghammer raises tame wolves. All the wolves there have been born in captivity for several generations, hand-raised as puppies, and "tamed." They were all part of Wolf Park's demonstrations and were handled daily. When Coppinger was taken into the cage with the socialized wolf pack, he was afraid. But Erich insisted. "Just treat them like dogs," he told him authoritatively. Coppinger complied "by thumping Cassi on her side and saying something like 'good wolf.' That was when she became all teeth. Not a nip, but a full war—a test of my ability to stay on my feet and respond to Erich's excited command, '*Get out, get out!* They'll *kill* you!' " Erich asked Coppinger later why he hit Cassi. He wasn't hitting her, he said; he was only patting her. But these were not dogs. They were wolves, wild even when tamed.

The dog colonies that consist of formerly owned dogs or dogs who have had earlier positive contacts with humans consist of sad, pathetic, often ill dogs. Their entire attention is directed, whether we know it or not, to humans. Dogs are no doubt eminently suited to physical independence, but they are psychologically ill equipped for being alone. They are too bonded to us. The Italian wolf researcher Luigi Boitani has demonstrated this in a series of scientific articles.

WHAT THE CAT has had to do to overcome a natural inborn tendency to be solitary is almost unfathomable and perhaps unprecedented, yet the instinct lingers in a certain wariness toward humans. I would go so far as to claim that the cat is the only animal whose innermost nature has changed to a large degree, from solitary to sociable. Cats are not an orphan species. Dogs might be described as an orphan species. That is, the entire species is dependent, emotionally and even physically, on bonding with humans. The term comes from attempts to explain the behavior of certain dolphins who are literally orphaned, and then spend their time seeking out humans and even choose to live near them. Examples are Jean-Louis, who lived off Cap Finistère in Brittany; Beaky, an old male who lived near the Isle of Man in western England; and Simo, a juvenile who turned up in Wales off Pembrokeshire. There are many others. Abandoned by their own kind, they seem to seek help and solace from us. Dogs are very much like this—not just individual dogs, but the entire species. But cats? Nah. They don't need us—we need them. They deign to associate with us, but they don't have to. It is a case of noblesse oblige. Wanting to help us? I don't think so. Individual cases can be found where cats offer profound comfort, but it is very individual and somewhat rare. Most of the time, cats watch us suffer in silence, even going so far as to yawn. I love them to bits, but let's be honest here. I have yet to meet a service cat. *Some* cats can be trained to do tricks (though they never see the point), but I would love to observe the methods used (you cannot successfully punish a cat). Will a cat comfort you when you are sad? He might, but I would not count on it.

Benjy by contrast lives to comfort. My friend Danny Goldstine, a psychotherapist from Berkeley, California, was visiting us in Auckland. As his wife Hilary and I were about to go for a walk on the beach with Benjy, Danny began to recount the story of the man who captured Adolf Eichmann, and how the man's entire family had been murdered in Auschwitz. While telling the story, Danny was overcome with emotion. Tears began to fall from his eyes, and he found it difficult to continue his narrative. Benjy ran over and urgently began to lick away his tears. There was absolutely no doubt in our minds that Benjy was offering comfort for what he saw as emotional suffering. Benjy might not know about the Holocaust, but he knows about suffering—he recognizes suffering when he sees it. His first reaction is to attempt to assuage suffering. Cats might feel your pain, but they will likely just saunter on by. We have not come to depend upon them for emotional solace the way we do with dogs.

HORSES OFFER AN even greater contrast: Horses are more or less indifferent to the human world. They don't really care if we are sad or happy apart from the impact our moods have on their own well-being. I believe horses are domesticated only to a point. They can go feral very quickly, much like pigs, goats, and cats. When I look at a beautiful large horse, I see a wild animal, one whose wildness has been contained, tamed, made to fit into our world, but an awesome creature not intended by nature for our use or our amusement. Even under the best of circumstances, a horse rarely gets to live the life horses were meant to live. Most of the time a horse is alone in a stall or in a small pasture with one or two other horses. Their focus is not on the humans who care for them. Their emotional lives are not dominated by thoughts of humans. Dogs, under the best of circumstances, are almost inseparable from their human companions, and the bond works both ways: The dog is as interested in people as he is in other dogs. The emotions of a dog are exercised constantly toward members of the canine species as well as the human species. This is why it feels to me intrinsically less cruel to live with a dog than with a horse. Horses are just not suited, by size, temperament, evolution, or their emotional constitution to be playmates for humans.

HORSES ARE KEPT in corrals or barns. Sheep are kept in pens. Pigs are kept in sties. Chickens are kept in coops. Many so-called farm animals are kept in cages. Goats are fenced in. Every animal on a farm is confined in some way, for the simple reason that otherwise they will leave. They are staying only because they are forced to stay. Not so dogs. We do not need to chain dogs at night (though we invariably confine them for their own safety). We do not force them to stay. They stay of their own accord because they want to be where we are. "Well, sure," you might say. "We are their pack." But just consider how complicated that is. No other animal considers a human a true member of the pack. Mostly we are feared; sometimes we are hated, and at best we are tolerated. But we are never allowed full membership into the pack or herd. We remain alien outsiders. Except to dogs.

ONLY TWO ANIMALS choose, often, to sleep with us. There is something very touching about watching the untroubled sleep of a dog or cat next to his or her chosen human. True, they can be awake and alert and ready for flight in an instant, but the very fact that they allow us to be next to them when they are at their most vulnerable is nothing short of a small miracle. That is when we know we are part of their family. (In cowboy movies, it is always at night, when the good guys are asleep, that the bad guys, mistakenly taken in on trust, reveal their true, evil selves.) The old cliché that if you lie down with dogs, you will wake up with fleas, seems most inappropriate here. When we lie down to sleep with our dogs, we wake up to their enthusiastic licking of our faces when they see us again in the morning.

Of the other animals that are popularly kept in homes, there are few people who would defend the rights of fishes in aquariums. Parrots, on the other hand, have their defenders—people are besotted with them. The claim is that parrots bond with humans in an even more intense way than dogs. This might be true, but it does not alter the fact that all parrots must be kept in cages and their wings must be clipped (more honestly described as a form of mutilation). Given their freedom, would any return? Perhaps, but most would opt for the wild. The close relationship parrots form with humans is more a caricature of the one they would form with a bird-mate. Since most parrots are strictly monogamous, when they are kept away from their own kind, they turn—for lack of a better choice—to humans for partnership. They are not friendly to all humans; quite the contrary, once they have chosen a mate, they will jealously defend him or her against all intruders. They do not generalize their love for *one* human to any other humans. We can be sure that dogs do not mistake us for dogs; they permit us to join their ranks *in spite* of who we are. Parrots, on the other hand, probably do believe we are some kind of exotic bird, or else they are so driven by their need for companionship that they give it no thought. Either way, they are not expressing a close relationship with us that entails any kind of species equality.

Pigs are a special case. It would require an entire book to do justice to pigs and especially to their potential relationship with us. I say "potential" deliberately, for few people have attempted to become close to a pig. Most human relationships with pigs are unfortunately limited to the disguised animal called "pork" that appears on their dinner plates. Because we eat pigs, we have been reluctant to recognize their intelligence, affectionate nature, cleanliness, and ability to bond with humans in ways that rivals the relationship with dogs. But, given half a chance, a pig will behave very much like a dog. A pig will come when you call him; he will sleep next to your bed if you permit him to; he will wag his tail when he sees you; he will accompany you on long walks through the forest and teach you new ways to appreciate what you see. In short, we have never permitted pigs to be the animals with us that they would like to be. This is a special crime for which our species will one day have to atone. Meanwhile we should at least stop insulting pigs in our daily use of inaccurate and libelous phrases about the nature of these marvelous animals.

WHAT I HAVE said about pigs is true of many of our domestic animals (goats and ducks, for example), and there are many books written demonstrating how close one can come to an individual domestic animal when that animal is treated with respect and kindness. But by and large, we have been reluctant to enter into any of these possible relationships almost certainly because of our own bad conscience. It is hard to feel intimate with an animal you eat. We don't eat dogs. A person who harms a dog in our society is considered a social outcast, a true pariah. The special place that dogs occupy in our psyche is a result of 40,000 years of history, but also of our own willingness to give another species the benefit of the doubt. It is not for their intelligence that we have selected dogs, but for their ability to love us. Loving us as they do has made us love them. The more they love us, the more human they seem to us; the more we love them, the more doglike we seem to them. The gap continues to close, and we come closer to true equality with dogs than we do even with members of our own species. Humans allow differences of language, race, religion, ethnicity, income, education, and social class to come between us in ways that we transcend when it comes to dogs. Dogs are pointing us in a new and better direction.

CHAPTER 5

ARE ALL DOGS WOLF CUBS?

Sometimes, when Benjy is stretched out on the floor and completely relaxed, I examine him, and often I see a wolf. He is about the same size as a wolf and the same weight. His teeth are not as powerful or as big as a wolf's, but they are still impressive: strong, large white canines that could (but never will) do considerable damage. Remember that even a 50-pound dog can exert a bite pressure of over 350 pounds per square inch, enough to take down a large human. On the other hand, a wolf has a bite pressure of more than *1,200* pounds per square inch!

Benjy doesn't resemble a wolf as much as some of the Northern breed dogs (Alaskan malamute, Siberian husky, Akita, Samoyed), but he is well muscled. Like a wolf, he has a desire to please, but in his case, he is as willing to please a lower ranking member of our pack (he is especially fond of Ilan) as he is the adults in our family. At least once he has howled like a wolf when I swam too far out to sea. When puppies become too frisky with him, he disciplines them the way a wolf does—with a short nip or an immobilizing paw to the body—but like a wolf he is very tolerant of the young. He also loves to lick our faces, not just to solicit food but also simply to express his delight in seeing us after even the shortest of absences. Like a wolf pup, he grovels too, moving his whole body in circles and finally rolling over on his back to expose his tummy. Unlike a wolf, though, he is not submitting—he is asking for a belly rub. I have not seen Benjy raise his hackles or growl as a warning, though most dogs do. He does not seem territorial like a wolf, and would probably lick an intruder to death, but in that way he is unlike most dogs. In other words, yes, Benjy is a wolf, but of course he is also not a wolf. He may have descended from a wolf, but he is also a dog and so he does things and behaves with our family in ways that a wolf would not. Unlike almost any other domesticated animal (including our beloved cats), when a dog displays zero aggression as Benjy does, he will remain *reliably* nonaggressive for life. In other words, these traits for which humans selected are now an integral part of the development of dogs.

Mind you, I cannot be absolutely certain how Benjy would react if somebody invaded our home who really intended harm. I like to think he would insist on being affectionate, but perhaps he would recognize the seriousness of the situation and surprise us. He does, from time to time, issue a very deep and loud single bark when a stranger walks up the outside steps of our deck. If I were approaching a house that had a dog who barked that way, I would not open the door. He has not shown a protective streak toward us, even the children, but perhaps he has never had occasion to do so. He *is* protective of Moko, one of our three cats. The other day on the beach, a dog began to chase Moko, who happened to be walking by. Benjy placed himself between Moko and the dog and decisively stopped the chase. He has done this three times now, so I know it is deliberate. What goes on in his mind? "Not this time, pal!" Moko and Benjy have a strange relationship: Moko loves to hide, and then when Benjy walks past, Moko leaps out and jumps in the air, arching his back like a fiend. It is a joke, but the best part is that Benjy knows it is a joke. He pretends to be scared, though, and takes off from the spot. Then he runs to the beach and begins racing at great speed in circles, with Moko looking on with satisfaction. He pretends to be awfully frightened and Moko pretends to be satisfied. Both display their sense of humor.

IN THE 1950S, the ethologist Konrad Lorenz noticed that adult dogs are more like wolf cubs than like adult wolves. This led some to raise the possibility that perhaps dogs are simply wolves in whom normal development is halted and thus retarded. This means they will have retained the characteristics of wolf cubs without going through the further stage of adulthood. There are various names for this idea: *neoteny* is the most popular, but there is also *juvenilization* and *paedomorphosis*, or the retention by adults in a species of traits previously seen only in juveniles. Usually what is seen as "retarded" is the somatic development, so dogs are generally smaller than wolves, with smaller teeth. It even affects their sexuality; while dogs are sexually mature at six months, wolves are not mature until two to four years old. (But this fact may say more about what humans expect and want from dogs than about their nature: We breed for this early maturity, and expect to completely control the dog's reproductive life.) Lorenz, much as he loved dogs, believed the wolf was a nobler creature. Dogs love us, according to this theory, because they are still puppies, not full adults.

And it is true that we can easily win over a wolf cub, but once the cub reaches maturity (including sexual maturity), we are no longer a parent, but a rival. Wolves at that point lose their respect or even affection for us and may challenge us, with disastrous results for both wolf and human. But Lorenz should not, in my view, think any less of dogs for retaining puplike behavior. Lorenz, of all people, should bear in mind that no dog ever joined the Nazi party! (Lorenz, as a mature man, did.) It is true that wolves have larger brains than dogs. But most of the size differential has to do with the cerebellum, the part of the brain used in sensory activity, especially balance. The size of the neocortex is about the same in both animals. The notion that dogs who have smaller brains than wolves are less intelligent is just not true; they are simply intelligent in different ways. This is equally true of humans. It is what we do with what we have, not how much we have.

Neoteny, carried to extremes, is almost a form of *fetalization*. That is, an adult dog (say a papillon or a Chihuahua) resembles the fetal wolf. But the word is now used primarily to discuss the retention of certain behaviors we associate with young cubs.

Scientists may first have conceived the notion of neoteny in general, by observing the Axolotl (a Nahuatl word of the Aztec people meaning *water doll* or, according to some, *water dog*). The Axolotl is an amphibian, a mole salamander, almost identical to the tiger salamander (*Ambystoma tigrinum*) found in the Western states and in Mexico. It is a fascinating creature for a number of reasons including its irresistible appearance and its ability to regenerate. But what caught most people's attention initially about the creature was the fact that the Axolotl exhibits neoteny. Ordinarily, amphibians undergo metamorphosis from egg to larva (the tadpole of a frog is a larva), and finally to adult form. The Axolotl, along with a number of other amphibians, remains in its larval form throughout its life. Nobody knows the reason why this happened. Random genetic mutation? It is considered a "backward" step in evolution because the Axolotl is descended from what were once terrestrial salamanders. But what caught the layperson's attention is how cute these animals look to humans. (Adult salamanders, with their protruding eyes, are not adorable to us.)

Many people misunderstand neoteny, even animal behavior specialists. Thus Temple Grandin, in her latest book, writes, "A dog never grows up mentally." On the face of it, this is nonsense. Dogs mature both physically and psychologically. I presume what she means is that a dog is never as mature as a wolf. But what does "mature" mean here? You cannot make these kinds of comparisons of an ancestor species with its descendants, and attempts to do so have been disastrous. It is true that most wolves have larger brains than most dogs, but it is also true that nobody knows the exact significance of this fact. Again, brains in humans have not evolved over the last several hundred thousand years. But it is not about size alone: Neanderthals had larger brains than *Homo sapiens*, but were not more intelligent. True, as noted above, wolves have larger brains. The claim has usually been that wolves needed these larger brains to deal with their complex relations with one another. But would anyone who knows dogs claim they have less complex social relations than wolves? Certainly not. The very fact that dogs must engage in complex relations with a member of an alien species, something that no wolf does, places an extraordinary burden of cognitive sophistication on dogs.

Consider Benjy at a crowded dentist's office, where he was

invited to come and lie on the rug. He is an instant child magnet, and at least five kids surrounded him, pulled various appendages, patted him in odd places, and rolled on top of him, yet he never lost his smile. He was in bliss. But he had to make fine calculations. After all, having complex social relations *with a member of another species* is something that dogs do much more easily and much more successfully than any wolf. This takes a lot of brainpower! Even at the level of dog meets dog, we can see a different kind of intelligence at work than the one used by wolves. Wolves meeting strange wolves mostly have one of two possible reactions: aggression or avoidance. Now it could be argued that a well-socialized dog also has a single reaction when he or she meets a strange dog: friendliness and interaction—precisely the opposite of wolves. Much will depend on the context, so the meeting can range from indifference to boundless enthusiasm, but the point is that dogs are making judgments on a fairly sophisticated level all the time. It cannot be explained away by saying it is simply hardwired.

Smiles, of course, are related to a sense of humor. Dogs are not the only animal to possess one. Goats have a sense of humor, and I have seen it operative in our cats too. Probably some dogs have a more developed one than others. The other day I called Benjy while looking off into the distance. When I turned around, there was Benjy waiting, looking up at me. Do I just imagine that he is grinning? I don't think so. He has a smile on his face, of that there can be no doubt. He gets the humor of the situation. He thinks it is funny that I don't know he is right next to me.

I love thinking about Benjy and the complexity of his emotional (and even cognitive) life. People too often take for granted what should really surprise us. As I think about Benjy,

I remember the late Claude Lévi-Strauss's famous saying, as it is usually cited: "Animals are good to think with." But what the famous anthropologist really said is: "*Il fait bon penser en compagnie d'un animal*," which means "it is good to think when you are in the company of animals"—somewhat different. When I am in Benjy's company, I begin to think of many things I would not consider outside his presence. Too often we accept as given what are nothing more than the accepted clichés of our culture. Why do some writers (George Steiner, for example) regard a dog's love for us as "an anthropomorphic conceit," but our love for them to be not only true, but for some, the deepest love they will ever experience? "What can be absolute is our love for the animal or animals in our lives, asking for no guaranteed return," Steiner says. Why not accept that Benjy's love for me is every bit as complicated and sophisticated as my love for him?

It would appear that we are programmed to like and find adorable any young small mammal. We *ooh* and *aah* over bear cubs, puppies, kittens, baby rabbits, and just about each and every other animal, at least for the first weeks of its life while it frolics and plays. The reason seems obvious: We are hardwired to like small, helpless animals with big eyes, a round head, small or no teeth, soft fur, and floppy ears. We find puppies cuddly. We want to hold them and coo over them and protect them, partly because they resemble newborn children. We have evolved to find our own young delightful so that we will protect them and raise them. All the evidence suggests that the great apes find their young adorable, which is one explanation for why sometimes a chimpanzee will attempt to kidnap the offspring of another chimpanzee. I don't see why it would be any different for other species, though I know of no research on this intriguing topic. Perhaps the word "adorable" is too anthropomorphic, and the phenomenon would be more easily researched if we referred to this with some other word—"protective," for example. There is no question that most mothers from any species will attempt to protect their young from harm. It would not matter whether the mothers found them "cute" or not. We are definitely in the realm of speculation when we wonder if we are the only species to think of the young of other species in this way. Is it at all possible that some animals look at baby humans and find them equally irresistible?

Various experts on domestication (the German scholar Helmut Hemmer in particular) have suggested that the features we find endearing are the very opposite of the ones that give a wild animal an advantage. Hemmer would explain the floppy ears of a puppy as signaling a *loss* of hearing. There is no evidence that wolves hear better than dogs. Not all differences are disadvantages. Dogs, for example, are color-blind, and wolves probably are as well. Dogs see the groups of yellow to green and red to orange and blue to purple, but they can't differentiate between members of each group. They are more sensitive to movement: They can detect a distant squirrel's movements before we can. Dogs evolved to be crepuscular; sensitivity to light is of greater importance to them than sensitivity to color. Humans need to see fruits and berries, so red and yellow stand out from the surrounding green foliage for us. Floppy ears or not, even in sleep, dogs' ears shift toward every noise, tuning out what is "normal" (e.g., your TV blaring) and listening for what is important (e.g., a stranger entering your home). Your dog can hear a sound four times as far away as you can. (Actually, that is just a ballpark figure; in fact, dogs can hear some sounds ten, twenty, or even a hundred times better than we can. It makes sense; wolves need to hear the tiny squeals of a mouse far more urgently than we do.) There is no evidence that dogs with straight ears, such as German shepherds, hear appreciably better than dogs with floppy ears, such as beagles, though there is a small difference. Because research with wolves—especially uncooperative wild wolves—is almost impossible to conduct with the same rigor we apply to supremely cooperative domestic dogs, we are not certain that dogs and wolves have comparably sophisticated abilities to detect the faintest odor. Dogs, we know, can detect a spoonful of sugar dissolved in enough water to fill two Olympic-size pools. The nose is as important to a dog as vision is to humans and provides comparably complex information.

WE OFTEN HAVE no idea how dogs know what they know. Smell? Hearing? Something else? When I visit a particular used bookstore in downtown Auckland, Benjy likes to stretch out in the shop's doorway. I watched him the other day as we both heard footsteps coming up the shop stairs. From where Benjy is positioned, he cannot see a person coming up the stairs; he can only hear the footfall or pick up the person's scent. Upon hearing the footsteps, Benjy did not get up, but his tail began to wag, and wagged harder and harder as the person (whom he has not yet seen) approaches. Benjy's increasingly wagging tail sounds like gunshot on the floor, but still he does not get up. How does Benjy know it is somebody he knows? When a stranger comes up the stairs, there is no tail a-thumping. So it must be the characteristic sound of a known person's footstep that Benjy recognizes. A pretty amazing ability, especially when you consider that he has met this person, my friend Richard DeGrandpre, an American writer now living in Auckland, less than half a dozen times. He likes him, but since laziness trumps amicability in Benjy, he confines his greeting to a loud tail thump, refusing to actually rise to his feet.

DO I BELIEVE dogs have telepathic power, like my friend Rupert Sheldrake, author of *Dogs That Know When Their Owners Are Coming Home*? No. Do I believe dogs have senses we do not? Yes. Take Benjy's assessment of dog-friendly people. All of us who live or have lived with cats know that cats take pleasure in looking around a room, selecting a hapless person who dislikes or fears cats, and then plopping themselves firmly onto the person's lap and settling in. Perverse, and we don't know why they do it. Dogs do the opposite—they seek out dog lovers as much as dog lovers seek out dogs.

We have selected neotenic features whenever possible in all domesticated animals. For a while in the 1990s, there was an American craze for Vietnamese pot-bellied pigs. People thought they were miniature pigs who would remain tiny their whole life. The fad died when people discovered that they grew into 150-pound adults, and the males with tusks—fun to be around because of their intelligence, no doubt, but a far cry from the miniature pigs that most fantasized owning. In terms of behavior, a wild boar is pretty much identical to a domestic pig (which means that to learn something about pig welfare, one must study the wild boar). For humans controlling the breeding of domesticated animals—sheep, goats, or cattle, for example—physical traits are most important. But we also select for behavioral traits such as lack of aggression. Unless you are breeding the *toro bravo*, or Spanish fighting bull, you do not want an animal that seeks to harm you. We breed for a short flight distance (cows should not run away from us), a lack of fear of humans (we need to approach sheep to shear them, though that is not as easy as some assume) and, perhaps most important, we want an animal who is willing to accept humans as the head of the herd. This is, of course, easiest to do with animals that naturally belong to herds and are used to hierarchy and subordination. Part of our fascination with cats is that they, pretty much alone among domesticated animals, are solitary by nature and accept us as head of household with great reluctance, if at all. Their motto seems to be the same as the Sun King (Louis XIV): *L'état, c'est moi,* or "the state, it is I."

In the case of the dog, we succeeded beyond our wildest imagination. We have "created" a creature who enjoys our company, and who wants to spend time with us even more than he wants to spend time with a member of his own species! This is neoteny in full flower. We can make the claim with some certainty that dogs are a neotenic species. But we could also make the same claim with respect to humans. We too have neotenic features, even physical ones: Born with no body hair, we develop relatively very little, even as adults. Human females develop even less. Like many humans, many dogs retain their juvenile curiosity, their desire to play, their lack of aggressiveness—indeed, their gentleness—throughout their entire life.

Successful neoteny involves features we rarely think about. Consider teasing. It is certainly a neotenic trait. But we cannot successfully or safely tease *any* domestic (let alone wild) animal, except dogs. You cannot tease a cat, at least not without getting into trouble. I am not talking about mean teasing, but gentle, playful teasing. Dogs know we are not serious. Why are they the only species to know? It would seem that the answer lies in our 15–40,000-year relationship.

The anxiety that children often feel at being separated from their parents, even at a relatively advanced age (say, seven), appears also to be a product of neoteny. Dogs too, even adult dogs, are prone to the same anxiety, so much so that many dogs are given antianxiety medication when their human companions go away for the day. (I would recommend finding a way to take your dog with you or finding a sitter companion for your dog instead.) The similarity between children and dogs is striking. I notice that when Benjy and I are walking up the hill to meet Leila, if I say, "There's Leila!" Benjy becomes frantic. "Where?! Where?!" He begins to race up the hill at breakneck speed. I often wonder why the urgency, why the seriousness of purpose, as if he were rescuing her from certain danger. I think it is because Benjy never approved of the (temporary) separation in the first place. When the whole family heads up the hill, and I lag behind, Benjy often stops and refuses to continue until I have caught up. He wants to make certain the whole pack stays together. Surely he is right.

Curiosity of the kind that we attribute to young children is also commonly found in dogs, even adult dogs. It is part of the plasticity or flexibility of young children, also a neotenic trait found in dogs. It is also possible that this ability to remain curious throughout our lives—another trait shared by dogs and humans—was mutually reinforced during our time together. Children love to explore *with* their dog, and it is the rare dog who is not willing to enter into this partnership.

Children may be one of the clues to the origins of this mutual dependence, for children's capacity for emotional bonds with animals is far more alive than that of adults. There is much to suggest that children's capacity for love is enormous, but that instead of improving it, we usually destroy it by various kinds of negligence, shunning, and abuse (physical, sexual, verbal, emotional). In response, children often direct their love to animals and in particular to dogs. Perhaps one reason dogs respond to children (Benjy certainly does) with greater enthusiasm than to adults is because hierarchy and rank, absent in children, can destroy or inhibit the kind of deep love that depends on mutuality and equality. You cannot have love that is coerced, paid for, or unequal (top-down). The corollary of this is that the love between a dog and a person who insists on being the strict alpha animal is not the same as the love that develops from equality. (This is my objection to Cesar Millan.)

THE INTEREST IN neoteny among humans has been well documented. Desmond Morris had already written about it in the 1960s in two popular books, *The Human Zoo* and *The Naked Ape*. But it was only when the Harvard paleontologist Stephen Jay Gould wrote his book *Ontogeny and Phylogeny* with the chapter "Retardation and Neoteny in Human Evolution" that the subject took off. When I spent some time with Gould at Harvard in 1999, he told me that he believed humans were a neotenic species, that we became the species we are precisely because of the advantages of neoteny. I had heard this years earlier when I was talking with the anthropologist Ashley Montagu (who influenced Gould) in Berkeley in the 1980s about the importance of breast-feeding (best until seven, he said, when the child's immune system is mature). He talked about how we speak with animals: in the same high-pitched tone we use for our own infants. He also told me something I have not seen in the literature, namely that one prominent trait of neoteny is the fact that young children love animals! This view of humans as neotenic seems to be widely shared now by many scientists. We value those exact same traits that we have bred for in dogs: We like to have children who are not aggressive, who are gentle, who enjoy our company, who do not fear us. We find these to be attractive qualities. In fact, it has been claimed that the reason men are attracted to some women is because they display neotenic features, even physical ones. Think about Vellus hair on some women. These are the short, fine, light-colored, barely visible blond hairs on the arms and back of many women that men often find so attractive. We also find women (and men too) who are particularly gentle and kind and compassionate to others particularly attractive—at least some men do.

Ashley Montagu explains the tie between love and neoteny, which seems particularly apt for dogs: "The infant's need for love is critical, and its satisfaction necessary if the infant is to grow and develop as a healthy human being. This need for love—that is to say, not alone the need to be loved but also the need to love others—despite all socially pathological deformative processes, remains the most powerful of the needs of human beings throughout their lives, and it is clearly a neotonous trait." Dogs and humans, then, are the ultimate neotenic species. No wonder they are attracted to one another.

HUMANS HAVE THE most extended childhood of any mammal (perhaps elephants and whales rival us here). A chimpanzee's brain is fully grown at about one year of life, whereas the human brain is not fully complete until about twenty-three or even later. No other animal gets its teeth as late as humans do. No other animal is still under the tutelage of its parents for so long (again, whales and elephants are perhaps an exception). I am tempted to say that no other species is as protective of their young as are humans, but I would be on shaky ground. Yet it is true that other animals whose young require years of nurturing before they can venture off on their own are eager to see their genetic investment continue on its path. My mistake in approaching a wild elephant was not because I had come too close to *her*, but because I had come too close to her baby. There are numerous accounts of whale mothers whose calves are harpooned attempting to destroy the ships from which the harpoons are launched. Obviously the investment of any animal who has just a few children is going to be greater than animals who have dozens (rabbits) or even hundreds (sea turtles). A sea turtle can afford to be indifferent to the fate of most of the young who hatch from her eggs, knowing that only some will survive into adulthood. An orangutan mother cannot allow herself to be sanguine in this way. Even bears, not noted for their emotional sensitivity as a species, can demonstrate a singular concern for their young.

But the mysteries of mother love are by no means plumbed yet. No other species seeks to instill juvenile characteristics into adulthood, but we appreciate the ability to be flexible (behavioral and psychological plasticity), remain curious, and to play in adulthood as much as we do in youth. We are practically alone in having almost no body hair. (Though I'm not sure of the significance of this except that it makes us dependent on outside agencies to keep safe and warm. Other animals can generate what they need from their own bodies.) No other animal our size has such small teeth that erupt so late in our life, which means we must not need them for protection. The late Wilton Krogman of the Center for Research in Child Growth wrote,

"Man has absolutely the most protracted period of infancy, childhood and juvenility of all forms of life, i.e., he is a neototonous or long-growing animal. Nearly thirty percent of his entire life-span is devoted to growing." Gould said that humans are predominantly learning animals, and it is our extended childhood that permits the transference of culture by education. We mature sexually far later than any other known animal (and this seems to be still evolving; women are having children later in their life than at any other time in history). Does sexual maturation lead to a certain rigidity? I think it does. No doubt this too permits the greater degree of flexible learning. We are striving to remain perpetual adolescents, not in the negative sense of refusing to mature, but in the positive sense of retaining those traits that allow us to remain young at heart.

There is a problem, however, with this theory when it comes to dogs: It has long been recognized and accepted as true that dogs have a rather narrow window of opportunity when it comes to socialization. Scott and Fuller in a seminal 1965 book, *Genetics and the Social Behaviour of the Dog,* argued that by about sixteen weeks, dogs can no longer be successfully socialized, or at least that the optimal period for socialization is between six to eight weeks and one to two weeks thereafter. Although there can be disagreement over the time span, I believe it is true that if dogs are kept in isolation, never handled, and never introduced to people and other animals, they will remain timid and afraid of others throughout the rest of their lives. Sixteen weeks may sound like a very small amount of time, but if we take into account the fact that dogs have shorter lives than we do (though nothing as short as wolves; in the wild, a wolf lives on average for only two years), and if we remember that children who are not exposed to other people *also* develop problems, it seems that the real problem is raising dogs or humans in an unnatural way and in an unnatural setting. Still, as a species, we do have a much wider window of opportunity for socialization than perhaps any other animal. In fact, this window might extend to the very end of our lives. At least this is what Konrad Lorenz believed: "Human exploratory inquisitive behavior—restricted in animals to a brief developmental phase—is extended to persist until the onset of senility." But so too is it with the ideally socialized dog: Dogs who are happy never lose their curiosity and their desire to explore the world with their best friends, as many excellent books from *The Hidden Life of Dogs* to *Merle's Door* have shown.

Benjy is now four. He has made it very clear to us that he is more interested in novel situations than in those with which he is already familiar. Sometimes, when we take him on a daily walk he knows well, he looks almost disappointed and walks at a slow pace, reluctant to veer off the path. But when we take him to an entirely new place, he is fully engaged, rushing about and showing quite clearly how pleased he is to be somewhere new. The best example of this was when we took him with us on a gondola. We were visiting the New Zealand town of Rotorua, which is famous for a luge at the top of a mountain. Our boys love going there. We didn't want to leave Benjy behind, so we explained to the gondola operator that he was a guide dog in training (well, he *was,* once upon a time) and that we wanted to familiarize him with a gondola. The operator was intrigued and allowed us to take him aboard. Benjy looked dubious at first, but once inside with the door closed, we began to climb, and Benjy walked over to the window and looked out. His tail began to wag furiously: He was delighted to be seeing a landscape for the first time from on high.

Juveniles, in all animals, are more playful than adults. They are less dangerous, tamer, slower, and weaker than adults. This is true of human children too. And it seems to be true across evolution as well. *Homo neanderthalis* was bigger, stronger, and altogether less gentle than the male *Homo sapiens*. It seems we are slowly evolving (even if it is not genetic) away from an earlier mode. We seem not to want to be killing machines, even if our history suggests we have been precisely this in the last hundred years (not that we had a sterling record before that, but the scale was not as vast as it has become). Even our admiration for speech and its concomitants, e.g., wit and humor, are examples. We no longer reward sheer strength, and even many women prefer men who are verbally dexterous to men who are tall and strong.

We admire neoteny when we see it in another species. We may be the only species that reacts to the babies of another species exactly as do its own parents, with a deep desire to protect and take care of the helpless creature in front of us. We are certainly the only species that brings another species into our den to feed it and have it live with us for no reason other than companionship (we are not ants raising aphids here), even if we originally did so for protection or as a hunting companion. But the very fact that we ascribe this ability to wolves, as we see in our legends of wolf children, demonstrates once again the similarity that we believe exists between the two species.

IT WOULD BE unfair not to recognize the downside to our love affair with the paedomorphic image of dogs. Once we were able to (say about two hundred years ago), we began to deliberately breed for dogs with a certain appearance (not for reliable behavioral traits), for what we thought to be the right look, *regardless of the effects on the health of the individual dog.* (The downside to the way we treat dogs really deserves a book of its own, and this does not seem like the right place to express my strongly held views on this subject.) Examples are well-known; perhaps the most extreme case of this process can be found in the English bulldog, once a powerful, athletic animal but recently described as the canine equivalent of a train wreck. With the odd head, distorted ears and tail, and ungainly movements, the bulldog more closely resembles a "veterinary rehabilitation project than a proud symbol of athletic strength or national resolve." Moreover, today most bulldogs are born by caesarean section, and they have terrible problems with their nasal and respiratory systems. They have such difficulty breathing during sleep that many of them die early from heart failure because of oxygen deprivation. These malformations are also found in other breeds that are brachycephalic (short-headed), such as pugs, Boston terriers, boxers, and Pekinese, and in those with stunted limbs, including the dachshund and basset hound. The German shepherd is a sad story. Because humans like the sloping in the rear—a completely unnatural position, but one that makes German shepherds look as if they are about to attack, hence seeming more dangerous—they bred for it, and the result is spinal malformations and lower back pain. We have deliberately induced a genetic disease to satisfy our vanity. It makes no sense. Or rather, it makes too much sense. If dogs bred us for the qualities they liked, we would all be as peaceful as Mahatma Gandhi or as playful as Robin Williams.

Though entirely subjective, some people find certain traits in dogs appealing, such as the flat faces of the bulldog or Pekinese, which leads to breathing problems because of set-back noses and shortened air passages; the low-slung eyelids of a bloodhound, which leads to chronic eye irritation; the wrinkled skin of the Chinese shar-pei that gives rise to skin irritation; the huge size of the Great Dane, which brings about hip dysplasia or malignant bone tumors because of the great weight on the bones and also an inability to cool down; or the shortened size of the legs of the dachshund, which leads to dislocated kneecaps and the inability to stay warm.

When humans attempt to take control of the process of neoteny, things can go very wrong, whereas if we simply leave it to evolution, we do much better. Now if humans are a neotenic species, why is only *one* other species equally neotenic, namely the dog? Once again, we get glimpses into the profound consequences of 40,000 years of two species sharing their daily life, their feelings, their homes, even their children. We have become more like dogs and dogs have become more like humans. For each species, it was a step in the right direction.

CHAPTER 6

THE FORTY-THOUSAND-YEAR ROMANCE BETWEEN HUMANS AND DOGS

When I wrote *Dogs Never Lie About Love*, I was exclusively interested in the love that dogs feel for us—far less in the love we feel for them and even less about the origins of love in humans and dogs. I took it as a given; I devoted all of ten pages to love in dogs because I assumed there was nothing new to say. But at the time I did not know about a new possible explanation for love in both species. Evidence that dogs and humans coevolved and that the domestication was mutual has given me an entirely different perspective. Several authors, among them David Paxton, Colin Groves, and Wolfgang Schleidt, have wondered implicitly or explicitly whether we may have coevolved with dogs. This raises many fascinating questions. Is it possible that the capacity dogs and humans each have for love is one dependent on the long-term relationship between the two? This long-term association must have far-reaching consequences, but nobody has yet attempted to determine what those are in any detail. I believe it is precisely the capacity for extreme mutual love that has been developed during this time.

As I pointed out in chapter 2, new research suggests that dogs have been living with humans for 100,000 years or more. While modern humans (*Homo sapiens sapiens*) evolved between 100,000 and 200,000 years ago, the cultural aspects of human behavior—bone carving for religious reasons, tools, ornamentation (bead jewelry and such), drawn images, arrowheads—only appear as a coherent package about 50,000 years ago. It is possible then that dogs and humans engaged in a kind of mutual domestication, that we evolved in tandem with each other, especially when it comes to certain kinds of cooperative behavior. That would explain the fact that of all animals, dogs alone have formed intense, close bonds with humans that border on mutual adoration, even something resembling religious awe.

THERE CAN BE no question that dogs are descended from wolves: The dog and the gray wolf have a genetic difference of only two-tenths of one percent (.2%). As Robert Wayne points out, our dogs are gray wolves, despite their diversity in size and proportion. But in spite of this genetic near-identity, wolves do not appear to have the same capacity for love that dogs have, at least with other species. Nobody has reported examples of wolves developing great friendships with members of any other species. Since no human has ever really succeeded in safely living with a wild wolf pack under natural conditions for any length of time, our knowledge of the limits of wolf behavior and emotions is not rich. Wolves raising humans is a cherished myth of our species, but I am afraid it is just that, a myth. There is no hard scientific evidence that a wild wolf ever raised a human child. Everyone who has ever raised a wolf or a wolf hybrid warns of the dangers: The affection can be great for a single human, but is rarely generalized to all humans, and tragedies abound. Even the most loved wolf can turn against the most loving owner, as is universally acknowledged.

I HAVE SAID that it is dangerous to proclaim some quality unique to our species yet completely absent from every other. Usually humans engage in this activity to demonstrate their superiority to other animals. We used to claim we were the only species to use tools, but Jane Goodall showed that chimpanzees did so as well. We thought we were the only animals with culture, but it turns out that almost all animals who live in groups have something resembling culture, something that can be transmitted from member to member, including the ability to wash food to remove dirt. Many scientists do not like to be reminded of the negative side of our exclusive abilities. We are probably the only animal who ever thought about genocide, and certainly the only animal who actually put the theory into practice, many times over. Dogs, who follow us willingly into most of our obsessions, have never developed an obsession with violence or with denying other members of our species their humanness.

Emmanuel Levinas, the French philosopher and survivor of the Holocaust, wrote in a famous essay of "the last Kantian in Nazi Germany": no other than an ordinary village dog. Levinas was serving in a French regiment when Germans in the Ardennes took him prisoner in 1940 during the Second World War. He was Jewish, but was protected from the death camps by being a captured French officer and was placed in *Fallingsbostel*, a labor camp for officers. Each day the French officers were marched through the outskirts of the city of Hannover, where they were treated with contempt and looked down upon as vermin, not even human. There was one exception: a stray dog who found his way into the camp. Each day, when the prisoners returned to their camp in the forest, the dog would greet the line of men with great excitement and friendliness. He was always delighted to see them. He was there in the morning when they were assembled and "was waiting for us as we returned, jumping up and down and barking in delight." Levinas notes, "For him, there was no doubt that we were men." To the dog, unlike the Germans, these prisoners were human—his friends. He wagged his tail each and every time he saw the prisoners. The prisoners named him Bobby to signal their hope that one day they would be freed by the Americans. But the Germans objected to this ignorant dog and took him away from the camp. That is why Levinas immortalized the dog later by calling him the last Kantian in Nazi Germany. This dog, like Immanuel Kant and like all other dogs, understood that humans are an end in themselves, and not a means to an end.

IT IS PROBABLY not safe to say that the human is the only animal who has developed religious feelings, for it is possible that dogs feel something akin to religious awe in our presence. Are we godlike creatures to them? The idea is not entirely absurd, for something mysterious is going on when it comes to dogs and humans. The same is true in the other direction: Humans are capable of deep and true love for dogs, love of the kind that many people believe can or should only be displayed toward other humans. Whatever our view of the rightness or wrongness of such love, I think it is undeniable that many people feel a love for their dog that is hardly different than the love they feel for those members of their own species to whom they are most close. When we read about the death of a much beloved dog, we often find ourselves with tears in our eyes, as many people did when they read about the death of Marley in John Grogan's best-selling *Marley & Me*. When Marley dies, Grogan's middle child, Connor, writes a note to place with the dog in his grave: "Through life or death, I will always love you. Your brother, Connor Richard Grogan." One of the best and most touching of all descriptions of dog death is to be found in "Through the Door," the last chapter of Ted Kerasote's fine book *Merle's Door: Lessons from a Freethinking Dog.* When his time had come, this remarkable dog held Ted's eyes, and the look in them said, "I don't want to be here." Ted put Merle under a clump of aspen trees, where they spent the next eight days—Merle lying on his side, Ted sitting beside him. Ted kept his eyes on Merle as Merle kept his eyes on Ted. Every so often, Ted would lie down before Merle and say, "Do you know you're the pup of my dreams?" And then he'd sing his song—*I know a dog . . .*—and Merle would chatter his teeth and exhale "ha!" when Ted reached his name. Then they would simply gaze at each other, their eyes saying, "I just want to keep looking at you."

We might wonder how this love for another species develops. Surprisingly it is often instantaneous. For the most part, this is not our experience with humans, except for the occasional "love at first sight." These experiences of spontaneous love with other humans are often doomed to disappointment, but this is simply not the case with dogs. How many people have ever "fallen out of love" with a dog? I have rarely ever heard anecdotes about such a thing. Mostly it is the other way around: We are sometimes slow to warm up to a dog but eventually succumb. Think of people who are not "dog people" but who inherit a dog, or whose children beg until they get one, and the nonbeliever slowly—and sometimes not so slowly—has a conversion. It can be the way they look at us, or the way they demonstrate who they are, but mostly it is the recognition of the unqualified and unconditional love they feel for us. Humans are always admonished to exhibit such love, but it is something we hear mostly from theologians, because in fact it is maddeningly difficult to achieve. Not so for dogs.

So much so that we humans can feel hurt. I know I do. What hurts my pride is the recognition that Benjy could very well transfer his affection to someone else with no great upheaval in his inner life. It is not just Benjy; this is probably true for most dogs. We don't like to think about it any more than we like to think about our human partner carrying on life with somebody new. Dogs need love and attention and affection, but I suspect it does not matter all that much who in particular bestows these things on a dog, as long as the attention is forthcoming. When away from us, do they pine for us, remember us constantly, sigh at our absence? Not, I believe, if they have a surrogate.

I saw this clearly when our family decided to return to the United States in March 2009. Because we were not certain the move would be permanent, we decided it would be best to leave Benjy in New Zealand for the time being. We would be in California for only two months and would then spend the summer in Europe. Where could we leave Benjy in California, since we could hardly take him to Europe? Bringing him to California was not problematic—New Zealand has no rabies and there are no quarantine requirements from New Zealand to the United States. The other way around, however, was difficult. Dogs coming to New Zealand had to spend one month in a quarantine kennel. This I could not imagine. As it turned out, we found the ideal dog-sitter—too ideal, in fact. Jessica Walker is a lecturer in animal behavior at the university in Auckland. She used to work for the Foundation of the Blind and had known Benjy there. Even then Jessica had a soft spot for Benjy. *He was so loving,* she explained. Indeed. So we arranged for her to house-sit; she could take care of our three cats as well.

She was there for six months, and it worked out beautifully. For in the end, we decided not to stay in California but to return to New Zealand. Jessica found the return difficult—Benjy had bonded with her as completely as he had bonded with us. I doubt he thought much about our absence (we will never know), but—and this is what compensates for what humans might be inclined to regard as fickleness—when we returned, he certainly remembered us. I came home a day before the rest of the family, and when I approached the house, Benjy was waiting for me. He rushed to greet me with unmistakable enthusiasm. But when Manu and Ilan returned the next day, I saw Benjy display nearly uncontrollable excitement. He could not contain himself from joy. Leaping in the air, licking without stop, squealing with delight—he could not get enough of the boys.

I have now realized this is simply a trait of the canine race. They love to distraction whoever they are with. Their need for love and affection is absolute, but as long as it is fulfilled, I don't think it matters to them who bestows it. It is not as if they make calculations based on shared interests, political ideology, or physical attractiveness, let alone such adventitious qualities as class or race. It doesn't even matter to a dog whether you are homeless or not. Witness the dogs that live on sidewalks, sighing with contentment to be lying next to their beloved person, financially solvent or not. Somebody made the argument to me that this speaks of their more simple minds. Perhaps it does. But is that such a bad thing? We don't admire somebody very clever who is also very cruel.

It could be objected (and has been, of course) that we really have no way of knowing for sure what is the nature of the feeling that dogs have for humans. Is it like our love for them? Is it like our love for our loved ones? Is it something else entirely? How can I assert that the love dogs feel for us is similar or perhaps even identical to the love we feel for them? Of course, this objection is unassailable. I cannot prove that when Benjy looks up at me with what appears to be adoration, he is feeling the same type of love that I feel for him or for my wife, Leila. But nor can I ever prove to the satisfaction of a scientific skeptic that the love Leila feels for me, of which I am absolutely sure is genuine, is the same or even similar to the love I feel for her. I am certain the two feelings are similar. All indications lead in this direction. But proof is lacking and always will be lacking when it comes to understanding love. There simply is no way now or probably ever that we can enter the subjective world of another human being or any other animal. Feelings cannot be quantified. But apart from this philosophical objection, most people have no problem recognizing love when they see it and in spite of the mistakes we make, the deceptions and self-deceptions we are capable of, we feel secure in our knowledge that all of us have felt love and that this feeling of love is similar from person to person.

IS IT NOT possible that dog love is merely a new example of anthropomorphism? That I am attributing to dogs the feelings

I have and of which Benjy is not capable? Of course there is such a thing as anthropomorphism, or projecting onto animals capacities they do not have. We do it all the time, especially those of us who live in close proximity to animals. Today (as every other day) I talked to my cats. I found Megala lurking deep in the tangle of bushes behind our house. "Ah," I told him, "at last I have found your hiding place." We all say such things to our cats and dogs, even though we know they do not understand what we are saying. We do not say *meow* or attempt to use words or sounds we believe they could possibly understand; we simply speak in our native language as if we were speaking to another person, using words that are meaningful to us, knowing perfectly well that they are meaningless to them. This is harmless anthropomorphism. It is not that I believe that Megala knows what I am talking about, though he may well understand the emotional tone of my voice. I am simply indulging myself. No doubt it reminds me of the literature from my childhood where all animals are spoken to and clearly speak back. Not even children, after a certain age, believe this is anything other than a useful or charming literary device.

What, exactly, did we learn from each other in our mutual emotional evolution? (And remember, it is not just learning, but dogs and humans encourage, inspire, reinforce, model, and mirror for one another—for dogs almost certainly have mirror-neurons, just like humans.)

✦ *Tolerance*. In the example we saw above from the late Emmanuel Levinas, this tolerance is often of a philosophical nature. Dogs have no prejudices. They are not concerned with human social status, and in fact do not recognize *any* of the human markers we use to separate other humans. For a dog, money, color, class, clothes, language, religion, or any other device that people use to mark themselves off from others, is simply irrelevant. There is no such thing as a badge for a dog (except, perhaps, the badge of a dog-catcher!). Tolerance is also sometimes literal for Benjy: He can withstand an infinite amount of "mishandling" from children because, I am convinced, he knows that the intention is good. He cannot be provoked by children poking him, pulling his ears and tail, or thumping him on his back. For him, these are all signs of affection and he takes them that way, banging his tail on the floor as if inhabited by a motor. (Benjy, though, is exceptional, and it is important to teach children to treat dogs gently, since many would not respond as benevolently as Benjy does.)

This lack of prejudice on the part of most (though not all) dogs is quite remarkable and worthy of emulation. When a dog meets another dog, it is immediately understood: *This is a dog, just like me.* There is no checking for size, religion, color, race, language, or any of the distinctions we make that can mean "less human." Dogs recognize all other dogs, regardless of qualities, as dogs. They are friendly and happy to meet and socialize and get to know one another. This is a kind of social intelligence our species has only intermittently.

✦ *Recognition and patience with differences.* Benjy certainly knows that the old people he visits with me each week on the dementia unit are different. I am not the ideal caregiver for my ninety-one-year-old mother. I become impatient. I want to leave. I want to get my visit over with as soon as possible. Not Benjy. He is happy to sit there, his head on the knee of a demented ninety-year-old-man who is thrilled by Benjy's touch. He is in no hurry to leave. When I see an old woman's grotesquely swollen feet, I feel ill; Benjy licks them. The smells of the ward's inhabitants put me off, but Benjy puts his face up close to theirs and passes out kisses freely.

✦ *A love for the new.* Scientists even have a fancy name for this: *neophilia*. We are easily bored. So are dogs. We love something brand-new in our environment. So do dogs. This is a good reason not to leave your dog at home alone. He or she needs companions, toys, novelty. The best course of action is to take your dog with you wherever you go as often as you can. Recent research has shown that dogs have an innate preference for the new: "These results support the hypothesis that dogs might be naturally predisposed towards neophilia."

✦ *Partnership*: Dogs take advantage of the unique gifts of another. When we point at an object, it seems that only dogs out of all animals understand, and look to see what we are pointing at. They recognize our signs for communication. Other animals, even when equally or more intelligent, do not understand this, and will not look in the direction we point, even if there is a buried reward there. We have learned to appreciate dogs' ability to smell and hear beyond our ranges, but they seem to know that our vision is superior to theirs. Even more important, they somehow recognize our language ability: We give commands, but dogs—only dogs—consistently obey them. (Try telling a cat "I said *stay*!") We call this obedience, but what is really astonishing about it is the recognition on the part of the dog that we are speaking to him and only him. We are not making random noises; we are *talking* to him. He knows this in a way that a wild animal does not. And he cares about this in a way that other domestic animals do not. This is more than simple obedience; it is part of the partnership. We have this partnership with only one domesticated animal on this planet. I would not call it obedience because it is, ideally, about pleasure and the joy that comes from working in tandem. This is beautifully expressed in Spencer Quinn's charming mystery *Dog on It*, where Chet the dog and Bernie the detective are practically twins.

✦ We have both learned a *lack of resentment*. Dogs do not experience *schadenfreude* (pleasure in the suffering of others) with each other. While there are humans who are sadistic or psychopathic, we have never identified dogs of this kind. Even fighting dogs—or rather, dogs who are forced to fight—do not feel the kind of hatred, or pleasure in the suffering of the other, that some humans feel for other humans. For humans who have been around dogs long enough, it is likely that dogs have had a civilizing effect on their emotions. Children who grow up to be sociopaths often begin by being cruel to animals, especially dogs. (Whether this is spontaneous or learned behavior from other adults is at present unknown.) They have not opened themselves up to the ability dogs have to make us gentler. Children raised with dogs, on the other hand, tend to be gentler with other animals as well. Growing up with dogs who are treated as members of the family makes children more tolerant of differences in other humans.

✦ *Tameness* could be seen as the ultimate domestic trait. L. N. Trut, from the Institute of Cytology and Genetics of the Russian Academy of Sciences where the famous "silver fox experiments" were conducted, has repeatedly explained that the "selection for tameness affects the genetic systems . . . at the level of the whole organism." She makes it clear that this one single trait, tameability, is what leads to all the other changes we find in these remarkable foxes. They began as purely wild foxes, and in the short span of forty years have come to resemble the most tractable of domestic dogs: They wag their tails, lick the hands of their handlers, and seek out humans for affection. The experiment began in 1959, and is the longest running experiment in history to study the behavioral genetics of domestication. An experimental population of silver foxes was selectively bred based on a single criterion: whether they could fearlessly and nonaggressively approach a human. Surprisingly, the domesticated foxes developed a higher frequency of floppy ears, short or curly tails, depigmentation of hair, and extended reproductive seasons, as well as changes in the size and shape of the crania and dentition. In other words, the behavioral changes were mirrored by changes in physiognomy of precisely the kind we find in the move from wolves to dogs. As Trut concludes in her latest summary: "This selection [of tameability] may be regarded as the key and universal mechanism of the evolutionary reorganization of animals during their historical domestication." What else is tameability but a synonym for gentleness? These foxes became that way by becoming attentive to the people who handle them, precisely as dogs have done over their domestication. We are able to see in this experiment a speeded up synopsis of the process of domestication of dogs from wolves, and what we see is the *influence of two species on one another.* What people have failed to recognize is that if wolves became tamer (read: more gentle), *so did humans.* While the scientists who conducted the experiment are interested in pigmentation changes, reproductive strategies, and other physiological alterations, they have missed, it seems to me, the bigger picture of the extraordinary changes in behavior. (We might even ask whether these experiments changed the humans who engaged in them.) We know that the tamest foxes went to homes, but we know nothing of their further history. It would be fascinating to have a follow-up: How much like dogs were these foxes in the end?

✦ We engage in mutual *helping behavior.* We take our dogs to the vet (even if they don't understand why); they find us when we are lost in the snow or drag us from pounding surf. I devote a separate chapter of this book to a discussion of what dogs do for us, since it is so important to my thesis. What we do for them is sometimes more problematic: We use them in war to take messages, help find wounded soldiers, even engage the enemy. They do not always return. Some might object that neither do soldiers, but there is a difference: Soldiers make the distinction between their side and the "enemy" side. Dogs do not. For them it is all a game, even if one with unforeseen deadly consequences. For them, we are all the same; they do not recognize the human conventions of "our" side versus "their" side. We also devote enormous resources to keep dogs from untimely death. People make fun of how much money Americans spend on their dogs (by 2007, $40 billion is spent annually on pets in general, more than double what was spent in 1994). But the truth is that our willingness to take heroic measures to make the life of our dogs safer, healthier, or more comfortable is a measure of how much they have entered our hearts, and is nothing to be ashamed of. Surely it is the opposite we have to worry about. While there is cruelty to dogs (witness the popularity of illegal dog-fighting shows), it would be hard to find anyone in Western industrialized culture who openly approves of dog cruelty, and even in the dog-fighting world, handlers would claim (erroneously, of course) that they are simply allowing their dogs to engage in natural behavior.

✦ *Smiling.* I have said this before, but it bears repeating: Benjy's smile is something to behold. It cannot be missed; it is not a figment of my anthropomorphizing imagination. Nor is it a mere interpretation. When Benjy smiles, everyone knows it. Like a yawn, his smile is contagious, and all who see it begin to smile sympathetically. The smile involves his eyes too: His eyes shine with pleasure; his whole face develops that special relaxed look: mouth open, tongue hanging loosely, slight panting. The smiling response in dogs is worth a serious paper. It is like that of no other domesticated animal. The distinguished animal scientist Juliet Clutton-Brock has pointed out that the dog smile (sideways grin of the lips and muscles around the mouth) is an attempt to mimic the human expression of pleasure, and is never seen in wild wolves. (Though a submissive grin is found in wolves—so while dogs may well smile in their own way, it is not likely they learned it exclusively from us.) However, animal behaviorists might see smiling simply as a sign of submission in dogs. But in animal rights advocate and veterinarian Michael W. Fox's new book, there is a photograph of a malamute courting a wolf "where one can see the open-mouth happy play-face when she is hugged by the male, who wears a submissive grin." I have definitely seen Benjy smile when he is clearly *not* submitting or feeling intimidated but is simply particularly happy. The goofy grin that many dogs engage in when they are playing is not all in the imagination of their human companions. We smile at puppies and they smile back. Could their smiles have evolved in tandem? We find their play adorable, and they like that we do. So they attempt to amuse us and we laugh and they learn about laughter and fun. (My three adult cats love to wait behind a bush and ambush Benjy as he walks by. He pretends to be frightened and all are satisfied. Cats too are subject to neoteny, although no instance has ever come to light of a wild cat ambushing another carnivore for the sheer fun of it.)

Many of these similarities between dogs and humans could be summed up with the word "attachment." We attach ourselves to dogs in ways we do to no other animal, except the human animal. Similarly, in a mirror image of our attachment to them, dogs attach themselves to us. This is no accident. Evolutionary biologists might claim that this is merely an example of "evolutionary convergence," where two species reach the same result from radically different pathways of natural selection. Some scholars have even suggested that the human infant–parent attachment and the attachment behavior of dogs toward their owners is a case of the same functional analogy or evolutionary convergence and is therefore insignificant. But how could this be? No other animal exhibits such behavior. We can, as I have explained earlier, tame many animals. But you cannot force an animal to love you. A tame wolf is not attached in the same way that a dog is attached, and tragedy is not uncommon when people forget that a tame wolf is still a wild wolf. A horse may appear attached, but most people who live around horses will acknowledge that the attachment falls far short of what we observe between dogs and humans. Horses do not want to sleep next to us. They do not or cannot, except as foals, play with us. They rarely follow us for long forest walks. In fact, they require fences. We are unreliable with them: We sell them or trade them. We ride them and race them and otherwise exploit them, more or less consistently. Rarely do we hear of somebody who acquires a horse for the sheer pleasure of her company. While some percentage of dogs may be acquired only for their usefulness, the majority are not. Most dogs are adopted specifically to be companions and only secondarily to guard, herd, hunt, guide, heal, or provide therapy. A horse is rarely acquired merely as a companion. Consider too a word we use with respect to horses: We "break" them. This is the shorthand version of the same term we use for elephants: Humans "break the spirit" of the elephant. It is a horrible but accurate expression. In the case of horses, we must do something similar, for horses would never otherwise allow predators to jump onto their ultrasensitive backs. In nature, big cats would leap upon the back of a horse to subdue and kill it, so a horse's natural instinct is to keep anything off this most vulnerable part of the body. Horses must overcome an enormous inhibition to allow themselves to be ridden.

We use a bit in the horse's mouth, something foreign, unnatural, and unpleasant. By easing up on the bit, the horse does our bidding. We "correct" horses by applying pressure with our legs, thighs, heels, or even spurs. The reward for the horse is not human pleasure (as with dogs—affection, petting, our presence) but to have the pressure relieved. And while they may be grateful once pressure is off, anything like spontaneous affection toward the one easing the pressure is rare, movies for teenage girls notwithstanding.

There are so many similarities between dogs and humans, two entirely different species, that it stretches credibility to believe it is entirely accidental, or that convergence is a sufficient explanation. Dogs and humans have evolved together and influenced one another, and are able to bring to one another the same kind of emotional depth they bring to their relations with members of their own species. It is not surprising that we read about great friendships between two dogs, or about the enduring love of one dog for another, just as we read about the same topics in humans. Perhaps these stories are there waiting to be discovered in other animals, but for the moment we only experience them in dogs and humans. It would appear that we are the only two species who have been able to experience such a complex array of feelings for one another. It is no small thing.

CHAPTER 7

WHEN DOGS BITE

"But surely," I can hear somebody object, "dogs can also be murderous, can they not?" They can, it is true. We have to remember that the dog is basically a wolf, and a wolf is a carnivore. Not only that, but the wolf is, by all accounts, probably the most successful carnivore ever designed by nature. A big cat hunts very well for a solitary animal. But cats cannot compare to a pack of wolves who coordinate their movements, using cunning more than stealth. Humans have evolved to fear carnivores. It could well be that in our evolutionary history we were often the victims of hungry, prowling carnivores (this was certainly true for the big cats). Hence the title of a new scientific book about human origins: *Man the Hunted.* Today, no one has any reason to fear a wolf. But was it always so? If we go back far enough into prehistory, wolves may well have presented a danger to defenseless humans. If we are to believe our own nightmares today, where ravaging wolves still figure, it is possible that we really were afraid of these animals, at least until they became our companions.

There are people today who still fear wolves, even if they have never seen one. There are even more people who fear dogs. Sometimes this is based on direct experience, but often it seems to be innate, a carryover from our earliest history. Given how many dogs occupy homes (and even beds) in America today, some seventy-five million, we have to assume that far more people in the United States love dogs than fear them. But because my main thesis is that dogs have influenced our very capacity to love, it is necessary to discuss in some detail the shadow side of dogs. At least we need to understand how it fits into the larger picture of dogs as a civilizing force in human nature.

DOG BITES ARE serious business, no doubt. The numbers are significant: Nearly four and a half million dog bites are reported every year. Bear in mind that most dog bites take place when a dog is tethered and feeling defenseless. (Is it surprising that people who keep dogs tied up in a yard all day often find themselves with frustrated, angry dogs?) We must remember too that nearly 40 percent of people who own dogs choose their particular dog because they fear crime and want a dog for protection. These people seek out large breeds commonly perceived as dangerous and aggressive—pit bulls, Rottweilers, German shepherds, chow chows, Akitas, Siberian huskies, Dobermans, mastiffs of various kinds, wolf hybrids, and exotic dogs such as the Fila Brasileiro, the muscular Argentine dogo, and the massive Japanese Tosa-Inu—and then inculcate the desired behavior in them. Are some dog breeds inherently more dangerous than others? It is difficult to say. What is clear is that male dogs are 6.2 times more likely to bite than female dogs, sexually intact dogs are 2.6 times more likely to bite than neutered dogs, and chained dogs are 2.8 times more likely to bite than unchained dogs.

ONE PARTICULARLY HORRIFYING example of dog violence toward a human is the story of Diane Whipple. Whipple was a thirty-three-year-old lacrosse coach (twice a member of the U.S. Women's Lacrosse World Cup team) living with her partner in an apartment in San Francisco. Across the hall lived two attorneys, Marjorie Knoller and her husband, Robert Noel. For reasons that remain murky and unexplained, they adopted a thirty-eight-year-old man, Paul Schneider, who was serving a life sentence in Pelican Bay State Prison. Schneider was a high-ranking member of the violent Aryan Brotherhood prison gang and was attempting to start a dog-fighting business from within the prison, using his own two dogs, who were both Presa Canarios (a breed from the Canary Islands) crossed with mastiff. These huge dogs (one weighed 140 pounds) somehow wound up living with the lawyers in their apartment building.

On January 26, 2001, Marjorie Knoller was home alone with the dogs when she decided to take them up to the roof. Whipple was just returning from the grocery store and was about to enter her apartment when one or both of the dogs suddenly and without warning attacked her. A neighbor heard her screams and cries for help. Whipple had seventy-seven wounds as a result of the attack and died shortly thereafter from loss of blood. In 2008, a court sentenced Marjorie Knoller to serve fifteen years to life for the death of Diane Whipple. It became clear that the dogs had been raised to be vicious.

IT IS HARD to talk about aggression in dogs without raising the question of pit bulls: Mark Derr, the well-respected writer about dogs whose judgment can usually be trusted, sums up the problem: "Pit bulls in all their variety are perhaps unique among dogs in their overall level of aggression, their reduced sensitivity to pain, and their inability to recognize the normal signs of surrender." My ex-wife has a lovely pit bull who is remarkably gentle with all humans, especially children. Much as she loves him, though, I notice she keeps a wary eye out for other dogs. He simply cannot be entirely trusted around them. People, yes; dogs, no. My wife, Leila, is a pediatrician, but as much as she loves dogs, she would never bring a small child into a house with a pit bull. Our fear of pit bulls could be ignorance, prejudice, or simply common sense, but I believe there is little doubt that pit bulls have some aggressive tendencies that are genetic and probably hardwired. Aggression must be innate since it is almost universally the case among pit bulls. (Of the 1.7 million dogs euthanized in 2008 in the United States, 58 percent were pit bull–type dogs.) Pit bull rescue organizations and advocacy groups suggest that if you are walking a pit bull you should carry a "break stick" so that if your dog bites a person or another dog, you can lever the dog's jaw open. Not very reassuring! The pet retailer PetSmart does not allow pit bull–type dogs into their doggy day-care programs.

That said, much depends on what a dog has learned and how he or she has been raised and socialized. I am sure there are some pit bulls so gentle that they would never dream of fighting another dog or hurting a cat or human child. Some are even used as therapy dogs (Mary Tyler Moore trained her pit bull to be a diabetic-hypoglycemia alert dog). In rural France one often sees a sign on a house: CHIEN MÉCHANT, which means dangerous dog. But we should remember the old French aphorism (nobody knows where it first originated): "*Cet animal est très méchant; quand on l'attaque, il se défend,*" which translates as *this animal is very malicious; when he is attacked, he defends himself* (more elegant, I admit, in French).

Last year the pro football player Michael Vick pleaded guilty to dog fighting. He kept fifty pit bulls on his fifteen-acre property in rural Surry County, Virginia. The dogs were chained to car axles. Those that "showed no taste for blood" when put in the ring were eliminated—beaten, shot, hanged, electrocuted, drowned, and in at least one case, slammed over and over against the ground until he died from the trauma. It was widely believed that no dog could survive such treatment with his or her character intact. "Ticking time bombs" is how they were described. The veterinarians called in were ready to euthanize the surviving dogs, and even some animal rights organizations such as PETA concurred with the recommendation. But they were wrong, as a story in the *Washington Post* makes clear. One former Vick pitty, now named Leo, is a mere year later so gentle and reliable that he visits cancer patients as a therapy dog in California. Teddles takes his orders from a two-year-old tot. Gracie lives with four cats she adores and sleeps with her best friends, four other dogs. Another, Jonny, is a certified dog-listener at Paws for Tales, patiently listening to children who are nervous about reading in public read aloud. Several more are certified therapy dogs (like our Benjy) who visit cancer wards. So at least in the Vick case—which is a pretty severe test case—we can say that no, a dog is not defined by being a pit bull. The early experiences of the Vick dogs were horrendous, but not definitive. The genetics was ominous, but could be changed. The potential for love must be there from the start. It must come from being a dog. It is just there. And it is especially there when we, that other species, are at hand to nurture it and encourage it and love back in response to it. "Of all dogs," says Dr. Frank McMillan, the director of well-being studies at Best Friends Animal Society, a 33,000-acre animal sanctuary in southern Utah, "pit bulls possess the single greatest ability to bond with people." I am not sure I would single out any dog breed as having a greater ability to bond with people, but I am certain that pit bulls do not have any *less* of an ability to engage in this bond. So when PETA and others felt *all* the Vick dogs had to be euthanized, they were falling into a trap: the false belief that while wolves can become dogs, once a dog becomes a wolf again (speaking metaphorically), there is no way back. When we read accounts of these very dogs, we see that they were every bit as traumatized as humans would be if put into a situation where death was likely. It would take a human a long time to recover too; that is why we use the word "survivor" for those who endure trauma and why we have such diagnoses as post-traumatic stress disorder. Michael Vick's dogs were not killers—the vicious humans who made them fight attempted to turn them into killers, and the result was that these dogs were terrified and hurt in their very essence. That some of them bounced back to inhabit their true nature shows us once again what love can do.

Writing for *Sports Illustrated*, Jim Gorant described one of the former fighting dogs in his new incarnation:

> *Zippy is not a big dog, but she's a pit bull, one of the Vick pit bulls, and she's up on her hind legs straining against the collar, her front paws paddling the air like a child's arms in*

a swimming pool. The woman holding her back, Berenice Mora-Hernandez, is not big either, and as she digs in her heels, it's not clear who will win the tug-of-war. "Watch it!" she says to the visitors who stand frozen in her doorway. "Be careful. Sometimes she pees when she gets excited, and I don't want her to get you." And just like that Zippy whizzes on the floor. Twice.

Berenice's six-year-old daughter, Vanessa, disappears and returns with a few paper towels. The spill absorbed, Zippy is set free to jump up and lick and wag her hellos before she leads everyone into the family room, where Berenice's husband, Jesse, sits with the couple's five-week-old son, Francisco, and two other dogs, who rise in their pens and start barking. But Zippy has no interest in them. Instead she leaps onto the couch where Vanessa's nine-year-old sister, Eliana, is waiting. Vanessa joins them, and over the next fifteen minutes the two girls do everything possible to provoke an abused and neglected pit bull who's been rescued from a dog-fighting ring. They grab Zippy's face, yank her tail, roll on top of her, roll under her, pick her up, swing her around, stick their hands in her mouth. Eliana and Zippy end up nose to nose. The girl kisses the dog. The dog licks the girl's entire face.

The story of these pit bulls reveals clearly my thesis in this book: that dogs and humans were meant to display love for one another. It is part of what it means to be a dog and what it means to be human *precisely because* we evolved in tandem, and this perhaps unique and extraordinary capacity that both species have to feel and show love is the end result of this odd but glorious experiment in nature.

Human fatalities caused by dogs are only twenty or so per year in the United States. The human homicide rate varies, but is at least 800 times that of dogs (something like 16,000 per year)! Or consider the statistics of aggravated assault in the United States, the use of a deadly weapon with intent to cause grievous bodily harm: There are approximately a million instances every year, as well as half a million robberies every year. And if dogs kill us to the tune of twenty a year, we kill several million of them every year by having them euthanized in shelters.

An even more frightening statistic is that while 38 percent of households in the United States have one or more dog (while only 35 percent have children), dogs on average only last in a home for less than three years. That is, households "dispose" of the dogs in one way or another just as they become mature. There can be no doubt that humans inflict terrible suffering on dogs, from breeding them in the thousands of so-called puppy mills (more like satanic dungeons) to keeping them chained in a backyard all day with no companions and sometimes even without access to water. There is no end to the terrible things done to dogs, both in the past and continuing to the present.

BECAUSE DOGS DESCENDED from wolves, it may be worth thinking about dog attacks, bites, and aggression in the light of similar behavior in wolves. (I wonder if people who are frightened of dogs also have a fear of wolves even if they have no contact with them, and if so, from where would it derive?) Unfortunately, real and reliable information is scarce. Most people are intrigued by whether wolves kill humans. They would and they could, but mostly they don't. (*Why* they don't is a different question.) Douglas Smith, leader of the Yellowstone Wolf Project in Yellowstone National Park, tells me there have been no attacks of any kind by wolves on humans at Yellowstone from 1995, when they were first introduced, until 2009. This does not prevent some people from falsely viewing wolves as the ultimate predator of humans. This in itself would make an interesting psychological study. Does it go back to our evolutionary history when we would have been far more vulnerable to wolf attacks? Was it ever a genuine danger? In fact, in the entire twentieth century, there appears to have been no fatal attack by a wolf on a human in the United States or Canada, even though there are nearly 80,000 wolves in the United States (particularly Alaska) and Canada. (The key word here is "fatal"—wolves do attack people and often with intent to kill.)

In Norway, where apparently nobody has been killed or injured by a wolf for the past 200 years, half of the entire population responded to a recent questionnaire that they are "very much afraid of wolves." However, wolves are a definite danger in India, where several hundred children appear to have been killed by wolves. In eastern Uttar-Pradesh in 1996, a single wolf was found to be responsible for attacks on seventy-six children all under the age of ten, of which over fifty were fatal. The reasons for the high number of children killed in India are not entirely clear and may have little to do with natural wolf behavior. In most of the cases investigated, the wolves were stressed by a lack of natural prey. Children are three times as likely to be unescorted as livestock, wolves are habituated to humans, and families are so poor, and the government pays a large compensation to the families of the dead children, that neglect could be almost willful. Bruce Weide notes, "We do wolves a disservice if we strive to mold them into saints of the wild . . . Wolves don't care if they're your totem animal."

Russia is a special case. This is where most accounts of wolf attacks on defenseless humans in sleds originate. Many of the accounts are meant to be fiction, or are fantasies influenced by fiction (children on a troika attacked by wolves is a staple of Russian literature). Unfortunately, the information we have about such attacks is not reliable. What is certain and known is that from 1950 until 1954, an average of 50,000 wolves were killed annually in Russia. When the subject is carefully investigated, as it has been in Japan where wolves were entirely eradicated by 1905, we find that wolves are conceived as "demons" whose death it is our duty to ensure. Knowledge of the real lives of wolves is deliberately kept from the general public to make certain they share in the mindless hatred that has fueled their extermination. The erasure of knowledge about wolves produced a silence in Japan that is a warning to us all.

How often dogs bite other dogs, seriously or not, is not known. It is certainly a fact that a small minority of dogs and a larger number of humans can be aggressive, even murderously so. But this does not necessarily reflect on the ability of the remaining members of either species to express love.

Our lack of concern for the feelings of dogs has a long and distinguished pedigree in Western philosophy. No less a figure than Immanuel Kant wrote in 1780,

> *If a dog has served his master long and faithfully, his service, on the analogy of human service, deserves reward, and when the dog has grown too old to serve, his master ought to keep him until he dies. Such actions help to support us in our duties towards human beings, where they*

> *are bounden duties . . . If a man shoots his dog because the animal is no longer capable of service, he does not fail in his duty to the dog, for the dog cannot judge, but his act is inhuman and damages in himself that humanity which it is his duty to show towards mankind.*

So according to Kant, the only reason it is wrong to be cruel to a dog is because this encourages us to be cruel to other humans. Dogs themselves are of no account. (Note too that Kant is only interested in kindness toward an animal who has earned it by serving us: What he would say about a purely wild animal can only be imagined.) Schopenhauer made the appropriate response. He said he regarded what Kant wrote as revolting and abominable, because "Genuine morality is outraged by the proposition that beings devoid of reason (hence animals) are *things* and therefore should be treated merely as *means* that are not at the same time an *end* . . . Thus only for practice are we to have sympathy for animals, and they are, so to speak, the pathological phantom for the purpose of practicing sympathy for human beings." Cruelty and a lack of empathy are not something that can be so easily partitioned in this way. People who are cruel to people are also cruel to animals and vice versa. Our character is not dual in this fashion but is of one piece. The issue is cruelty per se, not cruelty toward any particular species. The same is generally true of dogs: When a dog is friendly, or superfriendly, as is the case with Benjy, he tends to apply this attitude indiscriminately and across the board. Kant should have recognized that dogs enhance our moral character rather than claim that we should be kind to dogs to improve our behavior toward other humans. Kant was unexceptional; most philosophers get this backwards.

Having said this, I must also plead for how important it is to recognize that just as dangerous dogs are *individual* dangerous dogs—each one with his or her own particular history, the full knowledge of which is essential to understanding any particular dog's behavior—it is equally important to understand that even the most mild and kindly of dogs is also an individual with the whims and quirks and hidden trauma that we all, dogs and humans, harbor inside us all the time.

I had a particularly revealing example of this recently. The Foundation of the Blind was hosting a walk through a large park with my children's school participating. The walk was led by Rob Matthews, a blind man who had set many world records for running and who had lived with several guide dogs. I was delighted to participate. I was sure that Benjy was in for the time of his life: a long walk in the park; Manu and his school companions; many other guide dogs, would-be guide dogs, and failed guide dogs. It made me happy just to picture his delight. I could not have been more mistaken. For the first time since I had known him, Benjy behaved as if we were taking him to a chamber of horrors. As soon as he saw the dogs, the trainers, the line of children, he went into a kind of moral panic. He sat down on the grass, faced away from the action, and refused to move. Denise, who had been one of his trainers, recognized him and came running up. Benjy gave her the most perfunctory of greetings. Not like him at all. She was equally puzzled, but said she would get him to walk. She made a valiant effort, making the walk sound like something that happens once in a lifetime. Benjy did not move.

By now he was attracting the attention of other dogs and other trainers. Other dogs were ignored. The trainers offered advice, but Benjy would not budge. He looked miserable. He would not make eye contact with me. If he had been a child, I would know that he would soon burst into tears. Finally he conceded to walk a few feet, but no sooner had he done so that he sat down and refused to move yet again. This went on for two hours, throughout the entire walk. We were soon left far behind. I pleaded, I cajoled, I bribed with treats. Nothing worked. In the end, I had to return to where my car was parked, and drive to where Benjy seemed glued to the grass. Only then did he get up and jump into the back of the station wagon. As soon as we departed he was himself. He wagged his tail, licked my hand, and would have dearly liked to explain what had happened. He only lacked the words. So I am left to speculate.

The next day, the consensus among those I asked was just what I surmised: Benjy remembered his days of humiliation and failure at the guide dog training camp and was afraid he was being sent back. He thought our time together had come to an end, just like it had in his previous homes, and he would be returned to his original place of shame. (Yes, of course I recognize this is pure speculation—but it does fit the behavior. I can think of no other interpretation.) My main point is that whatever Benjy was doing was particular to him and to his history. It could not have been predicted. Each and every dog has his or her own personality, history, collection of memories, fears, desires, and even trauma, whether internal or external. We must recognize aggressive dogs in the same way. Not a single one is without an illuminating history, if only we knew it.

I HOPE THAT I have not given the impression that Benjy is the perfect dog. (Is there such a thing? We don't speak of perfect persons, do we?) One flaw in particular intrigues me, because I just don't get it. I have written that dogs live to walk. There was, I thought, no greater pleasure. It was part of being a dog, doing what they do best and love most, being out and about in the world. I believed this was always the case, because I have lived with many dogs and I have seen the expression on their faces when the word *walk* is uttered. Comprehension at the deepest level! Nothing seems to make a dog happier than the knowledge that they are about to go for a walk. (I cannot refrain from mentioning that many philosophers claim that dogs do not anticipate the future. Oh? What are dogs thinking about, then, when their tails wag in anticipation if not the pleasures of the *future* walk?) I was sure I had discovered (well, *noticed* is probably a better word, since everyone on the planet who lives with dogs knows this too) the universal pleasure principle for dogs: walking with their closest friends. Well, I was wrong.

Now, Benjy is not the brightest of dogs when it comes to understanding human language and gestures, but he certainly understands the word *walk*. The minute he hears it, even when it is used in casual conversation about somebody other than him, he runs and hides. At first I could not understand what was going on here. How could all my hard-earned dog knowledge be turned so quickly on its ear? The truth only dawned slowly: Benjy hates to walk! Yet another reason he could not be a guide dog. He will do anything to avoid a walk. To get to the house, he must walk down a steep path; to get away from here, he has to walk back up that steep path (about a five-minute hike). Benjy won't do it, at least not of his own volition. You have to put him on a leash, and then he submits with that odd look dogs get when they are doing something they would rather not: head down, eyes downcast, as if to say, "*You* of all people doing *this* to *me*."

How could a dog not want to walk? I took him to the vet to assess his legs. Clean bill of health. I decided to find out more about his puppyhood, so I called the people who socialized him for the Foundation of the Blind for one year. "Ah, yes," they said. "Benjy and walking." So you know the problem, I asked them? "Oh yes!" Turns out, Benjy has always hated to walk. I am still trying to solve this mystery. Because it *is* mysterious. He loves to be on a beach, racing back and forth. He loves to go hiking in the forest. I am beginning to get it, or at least I think I get it: He wants to do things on his own initiative. It is not defiance, it is not stubbornness, and it is certainly not willfulness. He just likes to do things when he wants to do them and he is puzzled that he is expected to do things when *we* want to do them. This is not a philosophical issue with Benjy. He seems genuinely clueless why we would want it any other way. And you know what? I think he is slowly training me to see things his way. Live and let live, that's Benjy's philosophy, and I am thinking of taking it more seriously than I ever believed I would.

I NEED TO stop pretending (as do some dog psychologists) that we truly understand the canine mind. I took Benjy to a lecture at the university the other day about . . . Benjy! He was the star. I assured the students that when Benjy entered the room he would greet each one with enthusiasm. But Benjy would not enter the room. He stood at the door, looking up in the air with puzzled, genteel aloofness, refusing to come in. What happened? I don't have a clue. There were a lot of dog people in the room and, truth to tell, they had no idea, either. The famous American philosopher William James put it quite well when he wrote that "marvelous as may be the power of my dog to understand my moods, deathless as his affection and fidelity, his mental state is as unsolved a mystery to me as it was to my remotest ancestors." Touché!

We have (or at least *I* have) a propensity to seek psychological explanations for behavior that appears odd to us. We like to feel we have understood something when we can glimpse the roots in an earlier experience. Perhaps I have been unduly influenced by my training (since abandoned) as a Freudian analyst. Psychoanalysis, as I thought when I began my training, was all about trauma. No trauma, no neurosis, no analysis necessary. I have since altered my view of psychoanalysis, but I continue to believe in the causal power of trauma. So Benjy's abortive training still strikes a chord in me when I attempt to understand his psychology. I am as susceptible as the next person to the neat dramas of Cesar Millan, *The Dog Whisperer.* They are simple and clear and uncomplicated. But I don't really believe that Cesar, or I, or anyone else truly understands the psyche of the dog. Dogs retain, to me at least, a recalcitrant element of mystery. I truly thought I had solved the mystery of Benjy's stubborn behavior at the guide dog walk. But months later, while walking along the boardwalk, for no apparent reason, he suddenly sat down and refused to go any further. There was nothing bothering him, no danger, nothing at all unusual about the setting, and yet he sat and could not be persuaded to continue the walk or even move at all. It started to make me angry (he was thwarting my will!), but soon I was more curious. What could possibly be behind this sudden and intense refusal to move? I found no answer, but it made me realize how difficult, how very difficult it is to understand the mind and soul of any creature on this earth—man, woman, or dog. But I find it a wonderful thing that dogs are so strange and unknowable.

I HAVE COME to believe that every dog is an eccentric. In the nice sense of that word. Take Socrates. He is an old dog who lives with our friend Jennie Michelson in Sydney. He is appropriately named. He looks wise. He behaves in a wise fashion. But it is a wisdom known only to him. It is hidden wisdom. He seems to have no normal dog needs. After a long night, I get up and suggest we go for a walk. He looks sad and says no. He will not leave the house. I encourage him. He won't come. Jenny explains that he wants to be wherever there is the most action. With people he cares about. For the moment, that place is the house, so he won't leave it.

EVEN MORE ECCENTRIC is Tigger. Whenever I visit my favorite city in the world, Sydney, I go to see my friends Ondine and Dror. Dror is an academic researcher specializing in kangaroos (about which he is one of the world's leading experts). Ondine, with her genius father Brian Sherman, is the cofounder of Voiceless, Australasia's leading animal protection foundation, and is a passionate believer in the rights of animals. They live in a beach-town suburb of Sydney called Bondi. The first time we visited the family, I was introduced to their dog, Tigger. He looks a bit like a dark dingo, I thought when I first saw him. Little did I realize how right I was. Tigger was a dog like none I have ever known. He was and remains standoffish. Not just with me but with everyone. But look at the word I chose—in Tigger's mind, I am sure, he is not standoffish, but *dignified*. He is probably correct. Once on a visit, we wandered over to the main street of North Bondi to eat at a vegan restaurant. I was surprised to see when we left that Tigger did not exactly join us, nor did he *not* join us. He went his own way—that is, he took a route somewhat parallel to ours. I was surprised. Who allows their dog this kind of freedom these days? Elizabeth Marshall Thomas did so in her extraordinary book about dogs, *The Hidden Life of Dogs*, but there have been few souls hardy enough to follow her. Dror and Ondine would rather not follow her lead. But they have no choice. Tigger is more dingo than dog. He loves to come along, but on his own terms, at his own pace, in his own way. He takes a circuitous route; we just catch glimpses of him now and then. But when we arrive at the restaurant, there he is. He lies down a few feet away, not with us, but not *not* with us either. When we finish eating and head back, it is the same story: He leaves a few minutes later and we can see him tracing our path, more or less (less rather than more). But when we get home, he is already there waiting (or rather ostentatiously *not* waiting, close but not *too* close to the door). He keeps a distance from us, but it is a physical rather than an emotional distance. He wants to be associated with us; we are very much in his sight and of course in his mind. Just as he is in ours. An independent spirit? Is this the ideal we have been searching for in dogs (and humans too, by the way)? The peripheral vision that allows Tigger to see us is also a way of being. At an angle. I am intrigued and curious in the extreme. What kind of behavior is this? What kind of dog is this Tigger?

Dror and Ondine initially wanted another dog to keep their sweet year-old puppy, Winnie, company. They visited the local pound where Winnie locked eyes with Tigger, a skinny, jet-black, wild-eyed, nine-month-old mutt with no known origins. Ondine is delighted to tell me Tigger's story:

The elderly pound volunteer shook her head and warned us against him. "He climbs fences, he's been to four homes already and every person brought him back. No one could handle him. He is wild and I don't know what will happen to him." She pointed us instead toward the small, doe-eyed dogs to the left, but Winnie wouldn't budge. "Well, we'll just take them for a short walk together and see. Maybe they won't get on?" we said hopefully. Next thing, Winnie and the black mutt were howling, jumping, racing in circles, and licking each other's noses. It was such unusual behavior for Winnie, who was an averagely social dog, that we stood there, mouths gaping in disbelief. There was no way that we weren't taking this dog home with us.

Unusual behavior was becoming increasingly evident. Tigger didn't bark, he howled. He didn't obey the usual commands like "sit" or "stay." Dror and I had grown up with dogs our whole lives. I had learned training techniques as a teenager. For three years, we tried to train Tigger. As soon as he was off-leash, he would not come back to us. The most delicious, mouthwatering liver treats (that made Winnie swoon) made no difference. Harsh words sent him further away from us. Tricky catching techniques, where we would approach from different angles to corner him in a dead end, was to no avail. He was faster than the speed of light and seemingly cleverer than Dror and I put together. He knew before us when it was time to go home and would make plans to be uncatchable. We could never, ever take him off the leash and then catch him to put it back on. He would end up following us home, never moving more or less than about a block behind us. If we had driven to the park, he would not get in the car until we started the engine and started driving away. Then he would run after the car for a few meters, and we would open the door and he would jump in.

Walking by the beachside one evening, we acknowledged defeat. We have two choices, we realized. Either we can keep him on the leash all the time and really break his wild spirit. Or we could just let him run free. We knew that the dangers of letting him go were serious: potential car accidents, getting lost, ending up in the pound or with someone unkind, as well as the criticism (and fines) we would need to endure from the Council, public, family, and friends about his out of control behavior. We didn't feel like we had another choice. There was no way we were going to tie this wild animal up for the rest of his life so we could abide by the rules. From now on, Tigger was free.

Shortly after we had made our decision, Dror met up with someone he had been associated with in his environmental education work, an Aboriginal elder and healer, Uncle Max. Max took one look at Tigger and said, "He looks just like the black dingoes I used to know up in the Northern Territory. Is that where you got him from?" It was an "aha" moment. Black dingoes climb and howl (not bark) and are, of course, untrainable.

Our first apartment had a deadlock door. Unless bolted with a key, the heavy door would shut by itself, look locked, but is easily pushed open. Tigger did so, and would head down to the beach, a five-minute walk, hang out with the local dogs going for their walks in the park and generally take it easy. When we took him out in the evenings, local people would greet him and tell us stories about where he'd been during the day.

Tigger could tell who the dogcatchers were. There were a number of Council employees who would look out for dogs at the park next to the beach, responding to complaints and serving fines of $150 for offenders. Sydneysiders don't like dogs near the beach, offending their pristine community with unsavory canine behaviors. Tigger was at the beach every day, and no dogcatcher was ever able to lay their hands on him. They are only allowed to issue a fine if they catch the dog. We received many Council cards and notes under our door from catchers who would follow Tigger back home, frustrated at not being able to get close enough to him to make the catch.

When we decided to move to Israel, we made arrangements to take Tigger and Winnie with us. Tigger soon learned how to get around Tel Aviv, a noisy, crowded, traffic-jammed concrete city filled with smells of the markets of Middle Eastern spices and raw meats as well as the scent markings of hundreds of thousands of feral street cats.

One day, Dror and my father Brian went for a long bicycle ride, traveling about two hours across six-lane highways and up into the neighboring Arab village to explore their way through small alleyways and over ancient cobbled archways. Tigger ran behind them, as he often did, following Dror's bicycle. This time, distracted by the smells of the markets, Tigger lost track of them. Dror and Brian were greatly distressed imagining that there was no way that Tigger could find his way home in the foreign area after having made so many turns and circles in their wanderings. Dror and Brian returned quickly to our apartment, ready to call the local pound to notify them of the missing dog. Tigger was already lying peacefully on the front doorstep waiting for their return.

Today, Tigger has had many adventures. His once shiny black coat is now covered in gray, giving him a wizened look of respectability. He still doesn't come when called and can't be caught, despite his once-agile body now being cushioned with the accumulated weight of an old dog. Winnie has sadly passed on so Tigger is the elder of the house, teaching our new young Israeli street dog (and now Australian immigrant), Ketem, the ways of the world.

I always told people when asked about Tigger's unusual behavior and what appeared to be our irresponsible ways in keeping him in check, that Tigger was in fact not a dog but our "flatmate." I really felt that we shared our living space with him and he chose to be with us of his own accord.

Tigger never took a particular interest in my daughter, Jasmine, when she patted him, turning his back and walking away. However, my twin boys have special-needs and although they are three years old aren't yet walking or talking. Tigger's reaction has been out of character: Every evening he comes into their room when they are listening to their stories and waits for a lull in the book to put his nose in front of theirs, asking permission for a lick. My boys adore him, reach out their hands (which is challenging for them) to touch his fur, and open their mouths (I know, quite unhygienic) to let him know they are ready for a lick. Tigger, at the ripe old age of twelve, has never shown such consistent affection and love to an adult.

There are two reasons I find this story so fascinating and why I have provided it in such detail. One is because it gives substance to my belief that each and every dog is a complete individual, with his or her own unique story. We come to know them not as generic dogs or as animals but as persons, or at least as personalities. They have their quirks, oddities, and eccentricities, and they endear themselves to us as completely as any other members of our extended families. They are part of that extension. Tigger's story will be told down through the ages by all members of Dror and Ondine's family. When people gather at holidays or birthdays, somebody will inevitably say, "Remember the time when Tigger . . ." And younger family members will listen and, fascinated, will pass on the story to their own children. If we pay attention, surely all dogs are Tiggers. But no dog is exactly like Tigger.

The second reason I find this story so riveting is that it seems to me that Tigger is a living illustration of how we came to bond with wolves. He retains some of the wildness, some of the independence of the wolf. He lives at a distance. He even loves us at a distance. As we move, he moves. And although he lives in Dror and Ondine's house, he does so just barely. He leaves when he wants and comes when he wants. He finds his own food if he chooses. He likes to follow them as they go about their business, but does not want to be followed when he goes about *his* business. He really does seem as if he is a wolf in the process of becoming a dog. It is like watching in real-time the evolutionary history of wolf into dog.

SINCE I HAVE made the claim that dogs and humans have coevolved, there is the puzzle of just how it is that dogs have evolved to be less dangerous than humans, by and large. I consider myself to be a very mild, easygoing person who rarely gets angry. But if I compare myself to Benjy, he comes out ahead: He is *never* angry. He is much more enthusiastic than I am, especially with new people, and he is willing to lick the feet of old people. There are dogs like Benjy who seem to have experienced a total extinction of all aggressive impulses. They never get angry. We do not see this in wolves, so we have to believe it is a product of domestication.

Since humans and dogs coevolved, the question arises of why we do not find this same lack of aggression in humans. We find it in *some* humans, true, but in *most* dogs. On the whole, it would be safer to leave a child with his dog than with his father. In 1999, an estimated 3,244,000 children were reported to Child Protective Services (CPS) agencies as alleged victims of child maltreatment in the United States. Every ten days in England and Wales, one child is killed at the hands of a parent. An average of nearly four children die every day as a result of child abuse or neglect (1,400 in 2002) in the United States. We need only compare this with about twenty total dog-related fatalities a year! Is it possible, though, that people who live with very docile dogs themselves become docile humans? Surely our temperament affects our dogs; why should it not go the other way as well? Hanging out with Benjy all day is bound to be good for my character.

Several authors, in particular Raymond and Lorna Coppinger and Stephen Budiansky, have claimed that love, affection, and all other associated positive emotions between dogs and humans are relatively recent historical phenomena. They claim that dogs began their history with humans by serving a purely utilitarian function. But this they can only surmise. We have no means whatsoever of verifying it. The fact that in a village in Zanzibar "dogs were regarded the way we regard rats; an animal ubiquitously present, a potential vector of disease, a scavenger, and occasionally a thief, whose population needs to be culled from time to time," really tells us nothing about ancient hunter-gatherers (nor I suspect is it even universally true in Tanzania as a whole).

There is compelling counter evidence. There is the excavation at Ein Mallaha in the upper Jordan Valley of northern Israel, a Natufian burial site where the skeleton of a woman with her hand cupped over the chest of a five-month-old pup, just inches from her face, was found, about which I wrote in chapter 3. Davis and Valla, who excavated the site, write: "The puppy, unique among Natufian burials, offers proof that an affectionate rather than gastronomic relationship existed between it and the buried person, an addition to our knowledge of the way of life of Natufian hunter-gatherers." There is no way of knowing if it was a tame jackal pup or a dog (it was too small to be a wolf pup, explains Dr. Clutton-Brock, the world's leading authority on domestication), but what is clear is that the burial was quite deliberately aimed at showing an affectionate relation between the two.

IN SOME CULTURES (Muslim, Balinese, Chinese, Indian—with exceptions in all of these cultures, of course), dogs are despised, or at least not liked. I grew up in one such culture, Jewish. In a recent novel, *Indignation*, Philip Roth writes, "Growing up, I'd never known of a single household among my friends or my schoolmates or our family's friends where the parents were divorced or were drunks or, for that matter, owned a dog. I was raised to think all three repugnant. My mother could have stunned me more only if she'd told me she'd gone out and bought a Great Dane." I have been trying to puzzle out just why Jewish culture has such disregard for dogs. To some extent it is old: In the Hebrew Bible, "the dog had a bad reputation; he was undomesticated, wild, and ferocious," as the historian Shlomo Toperoff explains. Even the Talmud and Midrash were uncomplimentary to dogs: It was permissible to kill a "mad dog" (whatever that meant) even on the Sabbath. But there is another aspect to this repugnance that I attribute to the experience of many Jews during the Holocaust: German shepherds were used to terrify and sometimes kill Jews in concentration camps and ghettos.

Most fatal dog attacks come from dogs who have been trained as killers, or are neglected, abused, or ill-socialized. Merritt Clifton, who did an exhaustive survey of dog attacks, gives an authoritative opinion: "In the German shepherd mauling, killing, and maiming cases I have recorded, there have almost always been circumstances of duress: The dog was deranged from being kept alone on a chain for prolonged periods without human contact, was starving, was otherwise severely abused, was protecting puppies, or was part of a pack including other dangerous dogs. None of the German shepherd attacks have involved predatory behavior on the part of an otherwise healthy dog." But it is also possible to excite a normally placid dog into aggressive behavior through unexpected actions on the part of a human. Benjy loves me to stare into his eyes. But I would not try this on a strange dog, no matter how easygoing. There is no study of the German shepherds used by the Nazis—pity, for it would be illuminating. No doubt, as with the Schutzhund tests, the dogs were simply engaging in a kind of game.

There are also Jews who are mad about dogs—me, my ex-wife, my daughter Simone, my two sons, and many of the one million Jews living in New York City. Some people are more affected by tradition than others. My parents raised me with dogs, but my grandmother was appalled and called them *bichu*, which means something like "an unclean bug." I also heard the word *trayf* (nonkosher) a lot from her lips when she talked about dogs, though I am sure she was not contemplating eating them.

Needless to say, dogs are unaware of these religious distinctions as applied to humans. For dogs, people are people, and dogs are not. Some people claim that dogs recognize their own breeds: Many people with golden retrievers have told me their dog prefers other goldens, but I am skeptical. I asked my daughter, Simone, who works as a veterinarian technician, and she said that this was pure mythology. But even if we agree that dogs possess none of the prejudices that humans do including cultural biases against an entire species, we have nonetheless seen that some of what appear to be human prejudices and fears may be based on something else, something to do with an inherent nature of a particular dog breed. How valid is this? Temperament in dogs is no more or less real than temperament in humans. A great deal has been made out of the personal qualities of different breeds of dogs. While there is no doubt that dogs from a certain breed have certain tendencies (to retrieve or to herd, for example), we can no more pigeonhole dogs than we can humans. The inclination to do so for both is very great. I believe the temptation stems from the fact that once we do so, we feel we have a handle on a matter of some complexity. If we can simply attribute all difficulties to a mysterious "temperament" or an inborn or innate tendency to be a certain way, we may feel we have accomplished or understood something. But I don't think we have. I agree with the words of Jerome Kagan, from his book *Three Seductive Ideas:* "The distinction between what an entity is at the moment of observation and what it was in the past or what it might be at some future time is honored by all of the life sciences. Psychologists are unusual in their frequent failure to honor epigenetic stages. Psychological traits are not stable structures hidden deep in the person's core. Some qualities, like today's attire, can change tomorrow." Temperament, whether in a dog or in a presidential candidate, is elusive in the extreme, and its source no less so. Everyone has a different opinion. There are those who say it is there from birth and can only be altered with great difficulty. Others say we are almost completely the result of socialization and enculturation into a specific society at a specific time and place. Then there are those who believe we are never a single entity, but shift from day to day. Our temperament changes, in this view, with our different experiences, and so is something one day only to be something else the next.

Is temperament the same in dogs as it is in humans? We have no problem in talking about what easygoing natures Labrador retrievers have, but we no longer talk about the excitability of the Irish because it feels racist to make these kinds of blanket statements. Benjamin Hart, a former dean of the School of Veterinary Medicine at the University of California at Davis, profiled the behavior of fifty-six breeds of dogs, looking at thirteen traits such as dominance over owners, aggression to other dogs, territorial defense, snapping at children, destructiveness, excitability, demand for affection, and so on. He asked four groups of individuals with extensive experience with dogs—obedience judges, dog show judges, professional handlers of dogs, and veterinarians in small-animal practice—to rate the fifty-five most frequently registered breeds of the American Kennel Club, plus the Australian shepherd. It seems fair to say that the experts were often in agreement. We can, I think, talk about breed characteristics. Nonetheless, when one reads individual accounts of dogs in such excellent books as *Merle's Door*, *The Hidden Life of Dogs*, and *Nikki, the Story of a Dog*, or others like them, what stands out is the distinct *individuality* of each dog, regardless of breed. Moreover, many of these dogs are of no recognizable breed. They are mutts, mixtures, crossbreeds, or their original breed is unknown. They have lovable features not connected in any way with their breed, their ancestry, or their lineage. Just as there has been a decline in the number of books with titles such as Raphael Patai's *The Arab Mind* (or any other nationality), though they were once very popular, something similar is happening with dogs. I read these books about both dogs and humans, but I would not put too much faith in them.

I HAVE ALWAYS believed that humans take a great deal of time to get to know one another. I am always astonished to read of a terrorist, rapist, or murderer who is described by his next door neighbor as such a mild, friendly, kind person. Why would we assume that a superficial acquaintance could plumb the depths of another human being's soul, when we cannot even know our own wives or husbands fully? Assessing the personality of a puppy may be reliable in the short run, but can anyone truly claim that we know the adult dog by subjecting a puppy to a short test? Trust takes time to build. "Trust me," people often say, and we think, "Why should I?" It is no different with dogs.

Benjy has taken a long time to trust me. For reasons I can only guess, Benjy did not like to have his head petted from above (I realize that for many dogs this is a dominance matter, but Benjy is completely unfazed by questions of dominance). Slowly, very slowly, he has come to understand that I mean no patronizing gesture; it is merely my way of showing affection. So now he not only permits it, if it is not forthcoming, he will solicit it. In the beginning, when I took him out on the beach, I often feared he would simply keep going and not return. In fact, there were times he would take advantage of low tide, and simply head off down the beach, around the cliff, along the rocks, to the small seaside town of St. Heliers, about two miles from our house. He was *so* friendly that one human being was pretty much interchangeable with any other in his mind. Given his past, who could blame him if he wondered why he should show deep loyalty to one family, if shortly he would simply be moved to another? But recently Benjy has changed. He has figured out our early (5:30 A.M.) routine, and once it is finished he trots back to me to head home. It is "his" home now, with "his" family, and I think he has decided or realized or hoped that this time it will be permanent, that he will not be moving anytime soon. This realization has allowed parts of his personality to blossom that were no doubt there in larval form, but went relatively unexpressed—from fear, I believe. The morning cuddle is the most obvious example of Benjy's transformation. Sometimes I leave the house to walk up the hill and do errands, and Benjy does not want to make the steep climb. I let him stay behind. When I return, he is invariably lying on the path and when he sees me, he becomes delirious with joy. People love to witness it: "He looks so happy!" they invariably say. And he does. He has that look of pure happiness on his face that is unmistakable to humans. It speaks to us exactly as the look of happiness on a person. We know it instantly.

BENJY IS PARTICULARLY fond of children. He cannot pass a small child without wagging his tail, and if the child pats him, the tail wagging becomes frantic. Should he see the same child again, even minutes later, his pleasure is doubled. He smiles and smiles at them. Sometimes when we set out to walk up the hill and he clearly does not want to go, I need only tell him that we are going to look for Ilan or Manu, and he begins to race (I never deceive him; it seems like a dirty trick). When he catches sight of them in the morning, he is thrilled to death to see them again.

I have stressed the importance of individual history, and why one-size-fits-all "training" or therapy does not work for dogs or for humans. It has taken us many years to recognize the importance of childhood trauma in people and how a person experiences it; it is now time to accord it the same importance in dogs.

We once took Benjy to a beautiful region a few hours from Auckland called the Coromandel Peninsula. We wanted to say good-bye to a house we had lived in for a while, before we planned to move back to the States. As we were walking along a rural road, we saw a chocolate Lab who came bounding over to Benjy as if they knew each other. Long greeting, great play. Then we noticed that the collar on this dog was on much too tight—plus he was so thin, it looked as though he had not eaten for days. Convinced he was a stray, we were ready to take him home with us. Fortunately we had yet to visit our old house, but once we did, the Lab followed. Lo and behold, he lived with the new residents there. They told us that he had been living with a pig hunter who beat him mercilessly because he was gun-shy. After they adopted him, the dog would often run away. They had no idea where he ran to until they discovered that he ran to hide under their deck, and stayed there all day. In other words, he was a very sensitive dog and as a result had been traumatized. When he saw a man wearing rubber boots (like the pig hunter) he would go insane. Now imagine if we had taken him home, and not known that he suffered this experience in his early life—we would never have understood his behavior. Any dog taken from a pound—or really, anywhere at all—has led a life with consequences that you may never come to know.

Another example: We visited a family with a very rambunctious female three-year-old Labrador cross. She began barking furiously at Benjy, who was immediately and for the next two hours completely paralyzed with fear. He stood facing the wall and refused to look at her or anyone else for the entire time. What, in his past, would account for this? It couldn't be simply a case of temperament. The answer lies in actual experiences of which we as humans remain ignorant.

BUT APART FROM individual personality, an even more urgent question is in need of an answer: How, if humans and dogs evolved together with affection as the glue, could we now have whole societies—in China, India, Indonesia—where dogs are not loved? Even when these societies or cultures are otherwise changing in a mostly positive direction? Is it that culture trumps evolution? Is it possible that the feelings are there, but that they are unexpressed? Given half a chance, unfriendly dogs, or dogs who have had unpleasant experiences with humans, overcome this and become friendly. Humans are changing in those societies too: Consider how many pet dogs there are in China today. It is estimated that about 70 percent of rural households in China now keep dogs, though there is no precise data on that. We can imagine that many of the dogs are there because they are guarding property or in some other way being of service. There was a time when only aristocrats were allowed to own dogs in China. But after the Communists seized power in 1949, to have a dog in the house was considered a sign of bourgeois affectation, and the animals were hunted as pests. This year (2009) in Mouldin County, a Yunnan province in southwestern China, the local government slaughtered 50,000 dogs after three people died of rabies: "Dogs being walked were seized from their owners and beaten to death on the spot, the *Shanghai Daily* newspaper reported. Led by the county police chief, killing teams entered villages at night creating noise to get dogs barking, then beat the animals to death, the reports said."

David Paxton has suggested (in the 2000 article I cite above) that "It does not matter whether people liked dogs or not, it matters only that human home bases with evolving dogs in them tended to survive more frequently than those without dogs." This is an interesting hypothesis, but I am not sure there is much evidence to support it (and he does not provide any).

The essentials of our emotional connection with dogs have probably not altered in thousands of years. There is no reason to believe dogs were any less apt at reading our moods and being sensitive to them 10,000 years ago than they are today. If thousands of years ago there existed words in our vocabulary for "warm-hearted," then how could that expression not be applied to dogs? Then and now it is in the nature of dogs to want to make us feel better. It seems highly unlikely to me that this sentiment would not be returned by some humans at all times, though not by all people all the time. Dogs inspire emotions in us; they always have. Human emotions have not altered over the last 40,000 years. Why would it have been any different then?

WE MUST BE cautious in assuming we understand the attitudes of a remote civilization, even when we think we do. I said earlier that certain countries, such as India, have or had negative attitudes toward dogs. So I asked my old friend Robert Goldman, a professor and world authority on ancient India who often visits India, to report back to me what he found there about so-called "pariah dogs." He told me there are two theories on the name of the stray and largely feral dogs that are found everywhere in Indian cities and towns. In India, they tend to be called *pie dogs* or *pariah dogs*. The latter is from the Tamil *paraian*, the name of a so-called untouchable caste (now known as *Dalit* or backward caste), presumably because of the low status of members of this caste, the miserable conditions under which they live, and the way they seem to be shunned by those higher up in society, whether dog or human. The term *pie* could be either a contraction of *pariah* or from a Hindi term *pahi*, meaning "outside." Bob told me, "My observations from living in middle-class Delhi for the past few years is that the dogs there belonged, like the people, to three groups. At the top of the heap were the pampered and overfed breed dogs of the well-to-do. Cosseted by the families but, like them, serviced (walked, bathed, etc.) by the ubiquitous domestic help who can be seen morning and evening in the posh colonies walking their charges, leash in one hand and a stout stick in the other to beat off the local feral dogs. Next on the canine social ladder are the compound dogs and strays that are semi-adopted by local domestic servants, especially the security guards and *chowkidars* who are found outside almost every house. They are fed on scraps and are semi-domesticated and fairly friendly, but can pack sometimes and be dangerous. Last are the true *pie dogs*, miserable strays, feral, often visibly malnourished and ill with mange. Some are missing limbs from serious fights and traffic accidents. They lie around everywhere and are mostly inured to the intense human traffic, so one steps carefully around or over them, but they do present a danger of bites and possibly rabies. In the little park outside our place in Maharani Bagh, I would often see the scrawny 'outsider' dogs who infiltrated the area set upon and driven off by the locals, their servants and especially the elite dogs."

THE PARIAH DOG is often called a "primitive" dog (the writer Mark Derr suggests we use the word "basic" instead because it bears less cultural baggage), but the term is not meant to suggest anything moral. It simply means that these dogs, and others very similar to them (including the recently discovered Carolina dog from the southern United States, the basenji from Central Africa, the dingo from Australia, the New Guinea singing dog, the Telomian dog from Malaysia, Canaan dogs from Israel, and pariah dogs from Egypt), have certain features in common. In fact, some cynologists (dog experts) say that the reason these dogs from all over the world have such a similar appearance is that they demonstrate "long-term pariah morphotype" (LTPM). They also seem to have certain behavioral similarities—e.g., almost none of these dogs bark. Perhaps they are closer to wolves because humans have not genetically altered them—nobody is trying to develop their appearance or behavior to serve human ends. The question arises: What happens when we breed for the opposite, for neoteny? Certainly we get dogs who *look* different than their wolf prototypes. But how does this affect their behavior? Nobody knows dingoes well enough to answer whether they are capable of the same affection as Benjy. My own sense is that all dogs are capable of this affection if the conditions are ripe. Those conditions depend on humans. When humans reach out, dogs respond. I saw this very clearly in my visits to Bali, an Indonesian island where attitudes toward dogs are appalling.

I visited Bali twice with my family, and both times we were dismayed at the sight of the numerous stray dogs one sees on the streets of every village. Nobody knows for certain how many dogs there are in Bali, but there could be up to a million in a country with a population of 2.5 million people. Only a very small number of these dogs are "owned," or house companions, and those tend to belong to expatriates living in Bali. But it was not only the poor condition of these dogs that shocked us. Even more disturbing was the explanation we were invariably given from many people for why dogs were beaten, had stones or sticks thrown at them, or were even casually run over by cars that routinely did not stop: "You see, these dogs are simply reincarnated thieves. The reason they have come back to earth in this terrible condition is because in their last life, they stole, and this is their karmic punishment." The Balinese are mostly Hindu (unlike the rest of the population of Indonesia, who are primarily Muslim), but while the belief in reincarnation is of ancient Indian origin, the idea that dogs were thieves is not an idea I have ever heard in India or seen reference to in any classical text. It is a peculiar Balinese idea and I have no idea from where it may have originated or even how old it is. What I do know is that it is an idea with terrible consequences for dogs. They suffer miserably as a result of this insidious myth. Not only do the Balinese treat the dogs abominably, they also feel justified and even self-righteous in doing so. It is only fit, they say, that a thief should suffer, and by treating the dogs poorly people are only acting in accordance with the laws of karma. I tried to argue the point (since my own earlier incarnation included being professor of Sanskrit in the University of Toronto), but it was useless. They required no proof, no evidence, and brooked no challenge or counterargument; they *knew* that all dogs were thieves. "Even if they are dogs born in the United States?" I asked. "Ah, no—in that case, they are enjoying the benefits of their good karma, for in America dogs are treated well. So this proves that they were *not* thieves in their last life."

I wondered if the very few dogs we saw who were clearly loved and belonged to somebody were considered to have not been thieves by virtue of the life they were leading. It was a very convenient and very immoral view in my opinion. Almost none of the dogs we saw ever had any contact with humans that was positive. They were never touched or spoken to. They were only beaten and yelled at. Why, one wonders, would they stick around the villages? Was it only for the few scraps of food they were able to steal or find? That is a possibility. But I have another idea. I was not totally surprised to meet several foreigners who told me that when they lived in villages, they were able to "tame" these stray dogs, and in no time the dogs showed their gratitude by behaving just as we expect dogs to behave: with love and affection, wagging their tails and showing their true nature. It is as if the dogs were waiting for the one enlightened person who would recognize their need for friendship. They were waiting for the right moment to correct the human misunderstanding. They were not thieves. The only thing dogs steal is our hearts.

Another possible explanation for the Balinese attitude is that without recognizing it, these views are simply a reflection of the attitudes in the larger Muslim culture of Indonesia. By and large, Muslims avoid dogs, regarding them as unclean. Not many people keep them as family companions, although in places like Egypt that is changing. Whether the Koran disdains dogs is disputed. One later Muslim tradition attributed to the Prophet claims that a prostitute secured her place in heaven by saving the life of a dog dying of thirst in the desert. A similar passage is found in another ancient text, the Sahih Bukhari (vol. 3, book 40, verse 551), one of the six canonical *hadith* (sayings attributed to the Prophet) collections of Sunni Islam. They were compiled by Muhammad ibn Ismail al-Bukhari (810–870 AD) and are considered to be the most trusted of all the books repeating the words of the Prophet. After the Qur'an, it is regarded by most Sunni Muslims as the most holy of books. As cited on the Web site for the Center for Muslim Jewish Engagement at the University of Southern California, Abu Huraira narrated that the Prophet said, "While a man was walking he felt thirsty and went down a well, and drank water from it. On coming out of it, he saw a dog panting and eating mud because of excessive thirst. The man said, 'This [dog] is suffering from the same problem as that of mine.' So, he [went down the well], filled his shoe with water, caught hold of it with his teeth and climbed up and watered the dog."

MANY PEOPLE IN cultures where dogs are adored, coddled, and treated as members of the family believe that "other" cultures are not so enlightened. People especially believe this about ancient cultures. But it is difficult to know how any individual culture in the past responded to dogs. Information for many of them is lacking. I have not found, for example, much literature dealing with Native American attitudes toward the dogs who lived with them. So we cannot generalize about indigenous cultures, or the attitudes of hunter-gatherer societies. Moreover, when societies do put their attitudes to dogs in writing (as in ancient India or ancient Greece, as we have seen), we are often surprised at how intense and contemporary their attitudes seem to be—just like our own! We do, however, know something about Australian Aborigines and their relationship with dingoes. When I was in Australia, I would hear many horrifying stories about the attitudes of the Aborigines to dingoes, including the "fact" that they often break the legs of these wild dogs to use them as heating blankets on cold desert nights. But there is a completely different view of the attitudes of Aborigines to dingoes.

Since there is no other culture with a record of such a long cohabitation, it is worth learning a bit more about dingoes in Australia, because the Aborigines seem to have had the same thesis that I have in this book: that we are closely related to dogs. For the Yarralin of the Victoria River region, "there was a time when Dingoes and Humans were all one species. We have now gone in separate directions—one dog direction and one human direction—but we are sometimes like siblings, being descended from ancestral Dingoes."

Recent genetic research has shown only small mitochondrial differences in the dingoes, suggesting that all the dingoes alive today are descended from very few ancestors—possibly even just one breeding pair. From this same source we learn that dingoes have not been with Aborigines for the last 40,000 years, which is probably how long the Aborigines have been in Australia, but only for about 4,000. In fact, the Aborigines themselves recognize this. Laurie Corbett, an authority on dingoes, tells of how Aborigines of the Kundi-Djumindju tribe dance a reenactment of the arrival of dingoes in Australia. They show the dingoes running in great excitement up and down the deck of a ship and finally jumping overboard, swimming to shore, rolling in the sand, and shaking themselves dry: They have arrived! Early on people who saw this reenactment assumed the Aborigines were explaining how the Aboriginal ancestors brought the dingo to Australia. In fact, it is about the arrival of the dingo with visitors, not the original black settlers. These visitors, according to another Aboriginal story, were Baiini—yellow-skinned people who made regular visits to Australia (from Indonesia, no doubt).

In 1788, the First Fleet sailed into Botany Bay in New South Wales. Newton Fowell, midshipman and lieutenant aboard the *Sirius*, wrote: "Thus, was our first intercourse obtained with these Children of Nature [*sic*]—about 12 of the natives appeared the next morning on the shore opposite to the Supply. They had a dog with them, something of the fox species." They called these dogs "Tingo" (an Aboriginal word that means *tame*) and the seaman noted that they do not bark "like our dogs, but howl." Seafarers introduced the dingo (probably via Malaccan trading boats—Malacca is the capital city of the Malaysian state of Malacca), a primitive dog that evolved from a wolf, into Indonesia, Borneo, the Philippines, New Guinea, Madagascar, and other islands (Australia in particular) some 3,000–4,000 years ago. It is also possible they were brought by the Makassar (the provincial capital of south Sulawesi in Indonesia) trepangers, who collect sea cucumbers for a living.

Dingoes were soon appropriated by Australian Aborigines, who had arrived in Australia many thousands of years earlier. They became so valuable to Aborigine tribes that the dingo became one of their sacred totems. Why? I would maintain that it is not only because they were useful, but also because they were loved. The explorer Carl Lumholtz in 1884 reported that Aborigines in north Queensland reared dingo pups "with greater care than they bestow on their own children. The dingo is an important member of the family; it sleeps in the huts and gets plenty to eat, not only of meat but also of fruit. Its master never strikes it but merely threatens it. He caresses it like a child, eats the fleas off it and then kisses it on the snout. It never barks and hunts less wildly than our dogs, but very rapidly, frequently capturing the game on the run. Sometimes it refuses to go further and its owner has to carry it on his shoulders, a luxury of which it is very fond."

Laurie Corbett points out that wild dingoes can be tamed but not domesticated. "Should humans determine and selectively breed certain standards and characteristics for dingoes, they will cease to be dingoes. A domesticated 'dingo' is not a dingo but just another breed of dog." He says they do not make suitable pets or companion animals in most urban or rural situations and should not be owned by members of the public. I believe that the relationship of Aborigines in Australia with their dingoes probably provides us with the oldest snapshot of how early societies of humans regarded dogs.

LIKE ANY LONG romance, the one between dogs and humans has had its ups and downs. Dog-loving cultures like those in the United States and Europe seem to be in ascendancy, and even cultures that have harbored ill feelings toward dogs in the past seem to be coming around. Unfortunately, there are no written records for most of this long and complicated history. Cave paintings, sculptures, archaeological remains, and even cultural attitudes from diverse cultures tell us some things, but can hardly be considered the whole picture. No doubt in every culture and at every time there were people who adored dogs and others who did not. There were those who thought they completed their family and those who believed they had no place in a human home. (Even today, when I visit farms in the New Zealand countryside, I see working dogs who sleep outside and are *never* allowed inside the house, which is probably also not that uncommon on farms in the United States.) What seems certain is that dogs always retained the potential to be what some cultures now regard them: our best friends. I think that what accounts for this potential is the long relationship between us, no matter how dormant it could be in individual societies.

CHAPTER 8

WHAT THEY WON'T DO FOR US

For many years in the United States (from 1927 until the early 1970s), the only service dogs were guide dogs for the blind. Since the early 1970s, however, dogs have helped people with other disabilities and conditions: hearing impairment (dogs are trained to listen for doorbells, smoke alarms, telephones); multiple sclerosis (these dogs are called "balance dogs" and help people get up and down from beds and chairs, push elevator buttons, pick up the phone); spinal cord injury (dogs push wheelchairs, open and close doors, turn light switches on and off); diabetes (dogs can scent hypoglycemia); Parkinson's disease (dogs can break a freeze episode simply by gently touching the rigid muscles); and early-onset Alzheimer's disease (if the person becomes disoriented or confused and says "home," the dog leads the way back home). From the accounts I have read, on the whole, people with these disabilities and conditions feel their dogs are aware of the good they do, similar to people with guide dogs for the blind.

In the United States, there is a new program using dogs in prisons: The toughest prisoners, often hardened gang members, are given a puppy to take care of. The results have been extraordinary. Some prisoners will say things like "she's helped me a lot because she helped me find the man that I was before I came to prison." One has only to look at the photos of prisoners with their dogs from programs such as Pathways to Hope (www.pathwaystohope.org/prison) to see how valuable the connection is for both the prisoners and the dogs. For the dogs there are no such categories as "convict"—these are simply the humans they love. Often, the dogs at Pathways have been abandoned and/or abused. The prisoners train them to be assistance dogs, which gives them a deeper understanding both of their own abuse in the past and of the abuse to which they subject others. The prisoners come to see the value of an animal that can help people who have been harmed in a similar manner. I think we will see a great deal more of this in the future.

I am convinced that dogs who alert their human companions to an impending seizure are in fact aware of the responsibility of their gift (for it does seem to be a gift, since most dogs, including Benjy, cannot be trained to perform the task reliably). I have talked to people who have seizures and who live with seizure-alert dogs, and they—and I—believe that the dogs are indeed aware that they are protecting loved ones—from something disagreeable at least, an impending disaster at most. My friend Belinda Simpson, an occupational therapist who now helps brain-injured people to recover, is also involved in training therapy dogs and seizure-alert dogs. These special dogs have a gift or a talent for recognizing the onset of a seizure in humans, warning a sufferer by herding the person to a safe place (a chair, a couch) and standing by to make certain he or she does not self-injure during the seizure. Some professionals distinguish between dogs who can alert that a seizure is pending and those (far more common and easier to train) who provide assistance once the seizure is in progress, by fetching a telephone or even dialing 911 on their own. Belinda did not succeed in training Benjy. As she told me, "I felt it was unfair to give Benjy a life I knew he would hate." Benjy was calm and friendly and easygoing, but he did not like training of any kind. Belinda explained, "When we would go into a supermarket he would plant his feet and refuse to move. The same was true in clothing stores. He hated the public access side of training."

Belinda knows a great deal about seizures, both from her own personal experience with them and from the work she does. Since this is such an important topic, and there are so few firsthand reports of seizure-alert dogs, let me quote from an account she gave me in October 2008:

In 1996, I was very unwell with severe, uncontrolled epilepsy. I was having up to ten (sometimes more) seizures a day. I couldn't leave the house or do any of my activities of daily living. As a surprise Christmas gift in 1997, I was given Bradley, a ten-week-old golden retriever puppy. He changed my life. When Bradley was approximately five months old, he began acting strangely. He would sit and refuse to cross a road, circle around my legs (almost tripping me) until I would sit or lie down, rest his head on my leg to stop me from getting up, lick my hand repetitively, and more. An animal behaviorist who observed what he was doing explained it as alerting behavior, and we then linked it to my seizures. I believe that Bradley knew I was going to have a seizure and that I needed to sit or lie down so that I wouldn't hurt myself. Until we recognized what he was doing, he received no praise or reward for his alerting.

It was more the opposite. I would get annoyed at his obsessing and would tell him off for not obeying me when I told him to stop. He didn't have any motivation to continue alerting me except his concern. And to him, it was more important that he looked after me than avoid being told off for disobedience.

Bradley and I did go to service-dog school when he was one year old, to learn task commands and public access skills so that he could be a trained assistance dog and accompany me at all times. It was there that I learned about "intelligent disobedience," where Bradley would disobey my commands if it was for my well-being. But he knew when I was about to have a seizure and that I needed to be sitting or lying down so I didn't fall and hurt myself. I learned to reward him for alerting to seizures by thanking him quietly and sitting down somewhere safe. The praise came after the seizure. Bradley was prepared to put himself in reprimand's way for my benefit.

My seizures became less frequent and less severe once I learned that Bradley would be there to alert me to an oncoming seizure. I became independent again and was able to live a fulfilling life with him by my side. Knowing that Bradley was there for me, I became less stressed about my seizures, and about life in general. My whole life changed because of Bradley.

Theo was my second seizure-alert dog and learned much of what he knew from Brads. His first alerting experience on his own was when we were out with friends for a meal. Theo was sitting quietly under the table when he suddenly leaped out from underneath like something terrible had happened or was about to happen. He jumped all over me. When I tried to calm him down, he jumped on my friends as well. Embarrassed by the display, I finally took him out of the restaurant. But he pulled on the leash and refused to settle. We finally took him to the car. Once I was sitting in the car with the door shut, Theo settled down. My friends and I just laughed it off, thinking Theo was sick of eating out and wanted to go home, but in the middle of our joking about it, I had a big seizure.

Theo behaved as though something terrible had happened or was about to happen. He got no reward—on the contrary, he was instead told off for not behaving properly in public. I learned over time that Theo had big reactions when alerting to an oncoming seizure. I had to teach him that it was OK and that having a seizure wasn't as bad as he seemed to think it was. I learned to calm him down and to get to a safe place so that he could relax. He thought something terrible would happen if I had a seizure and needed to learn that it would all be OK.

Thank goodness my seizures had settled down by the time I got Theo and I was only having a couple of seizures a week. I don't think Theo could have handled much more than that!

How dogs (and it is not specific to breed, gender, or age) know that an epileptic seizure is imminent is still something of a mystery. Perhaps it is a person's body movement, a strange scent, or some other clue we are not privy to. (Most trainers believe the dogs' awareness is due to visual clues such as facial expression, posture, and general behavior rather than scent.)

Belinda has talked to several other people who are lucky enough to live with seizure-alert dogs and says that most of them report similar accounts. The dogs become anxious and unsettled before the seizure and will not settle down until their person is in a safe spot. Sometimes a dog will stare intently at their person, or they will whine, paw, bark, or do something out of character. Something about his world is changing, and something about the world of the human he cares about is changing as well. This upsets the dog and sets him about putting their world right. We see it as heroic; to the dog, it is simply doing a job. But I find it hard to believe that seizure-alert dogs do not understand the seriousness of their task, and I do not hesitate to use words like *compassion* and *altruism* for these acts. Here again is a similarity between humans and dogs not to be found in any other two species: We support humans who are sick, *and* dogs who are sick, just as dogs do. Elephants take care of disabled elephants, but nobody has ever seen an elephant take care of a disabled human. Even if you argue that this is purely something we train dogs to do, we have never succeeded in training any other animal to do it, and that in itself is remarkable and worth contemplating. When is the last time you heard of a service cat?

I should describe here one remarkable exception. *CBS Evening News* with Katie Couric reported recently about a remarkable and unlikely friendship. There is an elephant sanctuary in Hohenwald, Tennessee, where elephants can live out the rest of their lives free from exploitation. The elephants generally pair up and bond with another elephant. But one elephant, Tarra, an 8,700-pound Asian elephant, chose a different partner: Bella, a stray dog who also found a home at the sanctuary. Dogs mostly avoid the huge elephants, but Tarra and Bella became best friends, eating, sleeping, and playing together. The depth of their bond was only revealed very recently. Bella suffered a spinal cord injury and was no longer able to move her legs or even wag her tail. She lay all day in the sanctuary office. But Tarra, who has 2,700 acres on which to roam free, spent all her time waiting outside the sanctuary office for her friend to get better. When Bella was well enough to be brought down to see her elephant friend, her tail wagged furiously, and Tarra gently touched the dog with her trunk. Later, Tarra rubbed Bella's tummy with her enormous foot while Bella lay very still, enjoying the touch. You can see this on extraordinary footage taken at the sanctuary. Tarra is no longer a wild elephant, one could argue, and her choice of friends is constrained. True, but it tells us something about the nature of dogs that an elephant would choose, just like us, to befriend one. I especially liked Katie Couric's comment: "They harbor no fears, no secrets, no prejudices. Just two living creatures who somehow managed to look past their immense differences. Take a good look at this couple, America. Take a good look, world. If they can do it, what's our excuse?"

IS THIS UNIQUENESS because of coevolution? Surely dogs are not *forced* to service us. They want to. They get pleasure from it. Or do they? I believe they do, but there is an influential counteropinion, represented by the writer Paul Shepard in a series of influential books and articles. Shepard takes his cues from Konrad Lorenz, about whom he writes: "He loved dogs, but he knew them to be gross outlines of the wolf. With diminished brains and congenital defects, these abducted and enslaved forms are the mindless drabs of the sheep flock, the udder-dragging, hypertrophied cow, the psychopathic racehorse, and the infantilized dog who will age into a blasé touch-me bear, padding through the hospice wards until he has a breakdown and bites the next hand." On a radio show, Shepard called dogs "slaves" and was unprepared for the angry calls from listeners that resulted: "Their reaction was like that of certain pre–Civil War Southern plantation owners who could point with defiant compassion to their grateful, singing cotton-pickers. The truth is that pets are subject to their owner's will exactly as slaves. Yet the term *slaves* may not be suitable, since human slaves can be freed by political and social action. The goofies, congenitally damaged, cannot. If freed they die in the street or become feral liabilities."

These are strong words. In the case of domesticated farm animals, I could have some sympathy with his position (if, as a personal consequence, he were a vegan). Certainly we have done pigs, cows, sheep, goats, chickens, and ducks no favor in raising them, as we do, as "food animals" (not a term they would ever think of using about themselves). We deprive them of everything that would be natural for them in their own ecological niche, from searching for food to having offspring to forming intense bonds with other members of their species. And of course there are times when humans treat cats and dogs with the cruelty they inflict on other humans. But, under ideal circumstances, both cats and dogs can lead rich and fulfilling lives, whereas it is questionable whether it is possible that any other domesticated animal can. I have argued elsewhere (in my book *The Pig Who Sang to the Moon*) that we can only appropriately use the word "happy" for animals who are living the life they evolved to live, under conditions that allow them the full panoply of their emotional repertoire. The very fact that all these animals are killed long before their natural life span is over demonstrates clearly that they are not on farms for their benefit, but for ours. There is no nice way of putting this, for we exploit these animals in every possible way: We take their babies and use their skin, their fur, their milk, their eggs, and their flesh. *Slavery* is too benign a term for what we do to these animals. We would do better to think of death camps.

BUT THE SITUATION with dogs is quite different. True, we often keep them on leashes, not something they would choose on their own. But they do choose, on their own, to share our beds, to take long walks with us, go swimming with us, ride along with us in the car, meet our friends, and play with our children. None of these activities are something they engage in grudgingly. They enjoy them. They want to do them even if they evolved to do none of them, at least with us. But we could argue that they do the equivalent of these things with members of their own species. So we could almost say that they "transfer" these behaviors onto us, much as a patient might transfer feelings onto his therapist.

Can we say the same about the services they perform for us that are not part of their natural heritage? Dogs obviously did not evolve to lead the blind, and nothing like this exists in the natural behavioral repertoire of wolves. But dogs do it, and they do not seem to do it with bad grace. This is, I recognize, dangerous territory. We can only guess what goes through a dog's mind when he or she is leading a blind person or performing some other service useful to humans. But it seems to me one explanation for why they allow themselves to be trained to do these things is the long relation we have with them. Thousands and thousands of years have resulted in this trust, this unique partnership. They give to us, constantly, and many of us give to them in ways we only give to other close family members.

IT IS NOT surprising that dogs can sense a seizure coming on. After all, we now know that they can smell melanoma tumors, and there is much recent talk of using them in diagnosing other cancers. In 2004, the ability to sniff out cancer was subjected to a scientific test. The conclusion was positive. The *New York Times* has recently summarized some of the new research and it is very promising. Benjy, as I mentioned, is a failed service dog: He failed to become a guide dog for the blind ("did not wish to work" was the final verdict the association gave him, because he was "too self-interested"), and also failed as a seizure-alert dog. This led me to wonder what was required of a service dog. Most important, what do the dogs themselves think of their work? Here is where I am most likely to fall into the swamp of anthropomorphism, because, I confess, it is tempting for me to see nobility in what the dogs do and to maintain that they recognize they are doing good works. Now, their interpretation of "good" may be quite different from ours in some cases, but in others, I suspect they overlap. It is true that dogs who sniff out drugs are not concerned with human categories. They are unlikely to think, "Aha, I caught the criminal!" Drug-sniffing dogs are merely playing a game, and the reward for them is not the triumph of justice but a simple snack. They are trained to find something using their superior sensory apparatus; they are not trained in the finer points of human forensics, and for the most part our ethical standards are irrelevant and opaque to dogs. We cannot expect them to make the same moral judgments about drug smuggling that we make. They make their own, of course. We know, from recent research, that they will cease to work if they perceive injustice in the form of one dog getting a larger reward for the same work than another one. Until this current research it was simply not known that dogs (like primates) were sensitive to perceived injustices of this kind. Of course, this is injustice in a dog's world. It would be much more difficult to train dogs to perceive *human* ideas of injustice. But it has probably never even been tried. When dogs are trained to fight for the United States Army, there is no moral dimension. It seems counterintuitive.

Certainly war dogs are not internalizing our classifications of "good" and "bad" wars. German war dogs did not fight because they considered Germans right and the Allies wrong, as opposed to the American war dogs who wanted to kill bad Germans. For all the dogs, morality did not enter into it. Much of what they did was entirely natural for dogs: During the Second World War, American soldiers could sleep soundly in the field, their dogs sleeping right next to them, for they knew they could never be ambushed without warning from the dogs. During Operation Desert Storm, dogs stationed in the Persian Gulf uncovered numerous booby traps set for American troops. An explosives-detecting Belgian Malinois named Carlo sniffed out 167 concealed explosives in two months. Although U.S. Army dogs saved thousands of American lives in the Vietnam War, when the soldiers returned in defeat to the United States, they either left the dogs behind or euthanized them on army orders. And then we wonder whether *dogs* are capable of gratitude. Clearly the troops did not recognize their own debt of gratitude to the dogs who served them so selflessly.

THE CLAIM BY most experts is that dogs who find people buried in rubble after explosions, earthquakes, or other disasters are carrying out what is for them just a game. The trainers maintain that the dogs are not aware of the danger facing the humans and are unaware of the fact that without their help, the humans could perish. Budiansky in his dog-debunking book *The Truth About Dogs*, writes: "Search-and-rescue dogs that find people in collapsed buildings have no more awareness that they are rescuing people than narcotics-sniffing dogs are aware that they are enforcing the nation's drug laws. As far as search-and-rescue dogs are concerned, they are playing fetch." I am not sure this is true, for dogs have a notorious sense of their own physical safety; they hate placing themselves in any kind of danger. Why is it so difficult to believe that they rescue people out of love for our species? They may need training to do it, but the motivation could still be affection.

If it is just a game for them, why do rescue dogs often look so worried or concerned? Why do they look so serious? And how do we explain rescuing behavior that arises spontaneously, as is often the case? How could dogs learn a game they have never been taught? Jon Katz, a fine writer about dogs, talks about one of his dogs, Izzy, in his latest book, *Izzy & Lenore*. Katz trained Izzy to be a hospice dog, to spend time with people who were dying. Izzy always seemed to know what people needed and was able to give it to them. Katz believes Izzy knew what he was doing: "After our hospice visits, I noticed Izzy seemed spent; sometimes he hardly moved for the rest of the day. It was tiring work for him, I could see, even if it wasn't physical. It took a lot of canine energy, in ways I couldn't fathom . . . I have no doubt that Izzy was healing me as well as helping the dying. Each visit settled me, grounded me, and rewarded me. Izzy surely would have picked up on that as well. He has a great heart; he lives to serve people."

Why do people so close to death relish a simple meeting with a dog? Katz believes it was a need for love without reservation: "Izzy simply loved them and accepted them, no matter what they looked like, smelled like, felt like." I have talked to a number of blind people who have guide dogs, and they all feel that the dog is doing something quite special, taking care of them because of who they are, because of the relationship they have established with a particular dog.

While I still believe this is often the case, I should also note that my own experience with Benjy has not led me to believe he correctly understands when I am in trouble. For example, not long ago I decided to walk home from about a mile down the beach when it began to pour. The tide was out and I thought it would be fine for me and Benjy to walk home along the shore. But as the heavy rain continued, the seaweed-covered rocks became very slippery and I slipped. Benjy walked on. I fell four times on that walk. Nasty falls. Each time it took me a bit to get back up, covered in mud and slime and soaking wet. Benjy turned around to look at me each time I fell, but paid me not the slightest attention. He just kept on walking, seemingly unconcerned. Even when I called him back, hoping he would come over, lick me, and show a bit of sympathy. Not that he could have helped. Still, I was upset that he seemed not to care. But that is only my interpretation at the moment, because I was feeling mortified by my constant falling and the lack of response from my best friend. It was clear to me, though, that Benjy did not understand. He simply could not evaluate my situation.

This is also the opinion of a recent scientific article, "Do Dogs Seek Help in an Emergency?" Two experiments—one in which people walking their dogs through a field pretended to have a heart attack; another in which a bookcase pins a person to the ground, causing the person to cry out in (simulated) pain and tell the dog to seek help—seemed to show that dogs did not understand the situations. The question "Do dogs recognize an emergency situation as such and intentionally take action to help a victim?" seems to have been answered in the negative. The experiment's constructors are aware of the obvious criticism: that since these "emergencies" were staged, it could well be that the dogs knew that the crises were not real, whether from the lack of pheromones or other cues used by dogs to evaluate danger. (In similar tests, however, humans readily interpreted the mock emergency as a real one.)

But then I thought, do we only love people or animals who are intelligent? I love the thought that there are different kinds of intelligence. Benjy certainly has emotional intelligence. His strength is not problem solving, but relational. (Problems he simply ignores—that is one strategy!) But suppose he didn't have emotional intelligence. Would I love him any the less? Is love contingent on such qualities as intelligence, sensitivity, courage, or any other positive quality one can think of?

WE HAVE ALL heard remarkable stories of dogs who put their own lives at risk by rescuing not just other dogs, but members of other species: humans, certainly, but also cats and a wide variety of other animals, both domesticated and wild. I am not talking about dogs who have been trained to do rescue work or who have learned to rescue, but dogs who spontaneously rescue just out of the goodness of their heart. At least that is what I think it is all about. They are not doing it for our sakes, because we are often not even there to witness the beginning of the rescue. They are not thinking about the possible reward, nor, bless their little hearts, about the dangers to themselves. Dogs risk their lives for others—this is unique behavior (ever heard of a cat who rescues birds?). I suspect that if I put out a request for stories of such dog rescues, I would be inundated. Just to test the waters, I asked a friend of mine, Sandra Kyle, who teaches languages at Massey University in Auckland but is also involved in rescuing parrots and other birds, to ask her friends if anybody knew such a story. Within a day I had one.

Lyn MacDonald has been doing bird rescue work for twenty-five years. She had turned her large yard into a refuge for sick, injured, and lost birds. As the years went by, she accepted more and more birds brought to her by the public, and soon her property was filled with them. When the garage and yard were overflowing with birds, she would bring them into the house. It wasn't unusual to visit her and find ducks recovering from botulism sitting in cardboard boxes on the living room floor. Employed as an animal welfare officer by day, she worked well into the night on her charges. About five years ago, a breeder dropped five Maltese terriers off at the animal welfare pound when no one was there. Lyn decided to foster one of the dogs—she called her Millie—and soon found out Millie was pregnant. In Lyn's yard was a small pool built for penguins to swim in. One day Lyn noticed Millie alongside the pool. "She must be drinking from it," Lyn thought as she went back inside. About ten minutes later Lyn came back out again. Millie was still in exactly the same position. "That's funny," Lyn thought, and went over to have a closer look. She could hardly believe what she saw: Millie was holding on to the back of a chicken that had fallen into the pool and was drowning. She had it by the feathers. "I could see in her eyes she was so relieved when I came," Lyn said, taking over the rescue effort and retrieving the chicken from the water (and eventually rehabilitating the near-hypothermic bird). "Another remarkable thing," Lyn said, "is that the chicken was the same size as Millie, and in holding its head above the water for so long, she had placed her own life in danger." There are, true, sporadic accounts of animals other than dogs who occasionally rescue animals from a different species. But they are exceedingly rare, and almost never does one hear about an animal who risks his or her life to save a member of a different species. But dogs do this all the time, or so we (me and most of my readers) firmly believe. Only one other species does this as well: humans. How do we account for this? I believe it is a kind of moral contagion: We infect dogs with our compassion and they return the infection in spades. It is a benevolent circle, one of the few found in nature.

DO GUIDE DOGS know that their human companion is blind? I would have thought so, but so far the research shows that evidently the dogs do not know. A parallel question is whether dogs know that other dogs are blind? My own experience suggests that they do not. Perhaps, in fact, they "know" it, but respond very differently than we would expect. One of the wolves at the Wolf Science Center in Vienna, Tayanita, is blind, but it is not clear if the other wolves are aware of this. Again, perhaps their concept of being "aware" is different than our own. As far as I can tell, however, the researchers there have not directly addressed the issue. Raymond Coppinger takes a very dim view of just about all service dogs. He certainly does not believe they are aware of their service: "They don't know why they are pulling a wheelchair, or what happens when they push a light switch." Later he asks, "What is being accomplished for the dog? . . . There is no evidence whatsoever that the dog is aware of what is being accomplished." His conclusion is bleak: "Of all the working dogs, service dogs have the most difficult jobs, and the least fun while doing them. Theirs is a stressful occupation with little reward. Many are simply sterilized workers, or slaves, with little to no biological benefit for performing well. Theirs is a dead-end occupation. Their symbiotic relationship with people is thus dulosis—slavery." Coming from Coppinger, however, a former sled-dog racing champion, this is a bit dubious.

One of the things that strikes me about Benjy is how willing he is to be led by the smallest child. It is not that he is unaware of their status as children; he just doesn't seem to care. He is simply not a hierarchical dog. My own sense is that we have paid entirely too much attention to notions of a pecking order. It is worth remembering that the original work from which almost all of our ideas about pecking order derive is a single article published in 1922! And it describes work that was done on chickens that were confined, hardly ideal circumstances to understand natural behavior.

Of course, service dogs have been trained to carry out service duties, but many of the people who train them think that the dogs enjoy what they are doing, not only because they are well-trained and proud to display their training, but because they believe the dogs have some sense, hard as it might be to categorize, of doing something important for the human with whom they have a powerful bond, something that the human is unable to perform for him- or herself.

When I take Benjy to the dementia unit of St. Elizabeth's Home for the Aged in Auckland, I am amazed by his behavior: He licks the feet of old people, kisses their hands, and sometimes licks their faces. But I am not yet convinced he is aware of their "condition." He may simply be expressing his exuberance in general when he sees people who are happy to see him. He does not kiss the hands of most people who come to our house, so he must recognize something special in hospital residents. He can do what I could not. I could not bring myself to even touch those swollen red feet. The smells repulse me. But Benjy is not the least bothered. I admire him for his selflessness. And yet Benjy could not be a guide dog. I keep repeating to myself the phrase they used, for it feels like a personal insult to me: "He had too much self-interest." I know they mean it in a quasi-technical sense, a term expressing a certain lack of drive in a dog. But it rings true to me in ways I find almost embarrassing. When I walk with Benjy, his agenda differs from mine. He pulls hard to smell something that interests him. He is not easily dissuaded. Perhaps I could "break" him of this habit, but I don't feel like it. Some trainers, noticing Benjy's habits at the walk with other guide dogs, commented that he needed "strong" commands. I suspect this is code for a more forceful, less gentle approach. It is a little bit like breaking his spirit. I don't want to do that. I admire Benjy's spirit. I love his mental independence, especially since it is combined with an unusual degree of affection. Nobody would doubt Benjy's deep love for humans. It is unmistakable. Why could it not be combined with an equally unmistakable self-interest? Maybe Benjy has his own ideas of self-respect. If we insist, he will place our interests before his own, but he would much rather live in a totally egalitarian world. Who can say that he does not have the more progressive view?

CONCLUSION

IT'S ALL ABOUT LOVE

People seem to have a fascination with uniqueness. We like to think about the fastest animal on earth, the strongest, tallest, heaviest. We also like to wonder about abilities. For years we believed that humans had various unique abilities that animals did not. Many of these beliefs have slowly eroded, but some remain. I have discussed a few in this book. As I come to the end of this book, though, one difference stands out for me. Benjy is only four, but he is probably almost halfway through his life. Every time I think about that, I can feel the beginnings of melancholy invading my system. I say that Benjy also sometimes has a sense of sadness about him. This is still true. But I am pretty sure he is not contemplating his demise or mine. After all, I am nearing seventy, so I know that Leila *does* think about this and it brings sadness to her. I don't think it is justified to use the word *melancholia* when speaking of dogs. When Benjy sits next to me on the beach on a quiet evening and we both stare out to sea, I don't say to him, "*sic transit gloria mundi*, pal." Life is fleeting, Benjy, for both of us (and the death of the Empress of Austria means squat to us both). Do you feel it? Only I can worry about the shortness of his life. And I do. I'm glad he can't imitate Leila. One in the family is enough. But Benjy, unlike some other dogs, does not insist on teaching me about the glories of living in the moment. He does trail his own past. He is used to disappointment, to the transitory nature of even the best relationship.

I think I know why Benjy did not make it as a guide dog. Among other things, it was because he could so easily become distracted. I was told that when he was in training, food, another dog, even a blade of grass could take his attention away from his work. But I sympathized with him. After all, who would want to be a slave, even a beloved slave, to a member of another species? Given the choice, I mean. Not that I believe Benjy knew what he was being trained for and voted with his paws. It is just that he goes for immediate pleasure above all things. The nice part is that it works both ways: He is not interested in immediate pleasure for his sake alone, but for yours too. An obvious example that I have written about here is when he meets somebody he knows well from the past. He cannot be restrained. He knows the command "down," but he cannot, literally *cannot*, obey such a command, as he is being guided by a far more powerful force: the desire to demonstrate his love and recognition. So up he bounds, kissing the person full on the face over and over. Our thirteen-year-old son Ilan was away in Hawaii for one month of summer camp. When he returned, Benjy could not contain himself, leaping into the air to reach Ilan's face and lick it. Such is Benjy's enthusiasm that it rarely fails to infect the second party. I have yet to see even the strictest dog trainer (and they do tend to be strict, don't they?) not succumb and revel in his attention. Because the love is so *obviously* genuine.

I am not suggesting that dog affection is any different in

other dogs, but Benjy is truly beyond the pale. He twirls, he groans, he moans, he leaps, and his eyes shine bright. How on earth could anyone resist? It is giving him such obvious joy that it brings joy to his victim (if being kissed on the mouth by an 80-pound Lab is indeed victimhood). And then he suddenly settles down with such a sigh of contentment that the meaning is unmistakable, even to those who do not speak dog: *All is well with the world. It can't get any better than this.* This is how Benjy lives his life: one day after another of blissful interactions. He is the canine love guru. Wherever he goes, he reminds people that dogs and people, and dogs with people and people with dogs—it's all about love. And the best thing is there is nothing phony about this guru; this is the genuine article. I have yet to meet a guru I did not dislike. But Benjy is different. I have always believed: If you want a real guru, get a dog!

Having a dog is like falling in love, except that it usually does not take as long to fall in love with a dog as it does with a human. But the similarities are striking: In the ideal relationship with another person, we become closer and closer as the love intensifies. The joy that dogs have in our presence grows as the familiarity grows, as does the trust and the love. It takes a while before a dog or a human can trust enough to show you the love of which he or she is capable. We need the right conditions in order to manifest love. And there are many things that can get in the way of love coming fully into play. This is why so many marriages fail. This is why we often *don't* see a dog displaying love. The conditions must be right. The atmosphere must be prepared. But when it happens, there is little that brings as much happiness to both parties.

Benjy did not fall in love right away. Neither, to be honest, did I. We both seemed like strangers to each other in the beginning. This changed for me just as it did for him. But it did not need to change for Manu and Ilan, because the love was there from the very start. Why is it that children feel less strange with a new dog than other family members do? Benjy was totally familiar to the boys right from day one. It is rare that we ever get the feeling our dog is judging us or that he finds us lacking. The mysterious hand wrote on the wall of Belshazzar's palace: *Mene, Mene, Tekel u-Pharsin* ("You have been weighed and found wanting"). These words are just not part of a dog's vocabulary.

"Look, he's grinning!" "No, he's smiling!" "No, he's laughing!" Benjy is in his favorite position (on his back) in his favorite place, the schoolroom: surrounded by twelve seven-year-olds all poking at him, patting him, pulling him, and lying on him. He revels in it. If they stop for even a second, he will put a paw on a child's arm as if to say, "Please don't stop what you were just doing." His tail thumps the wooden floor hard, beating time in a steady rhythm.

The kids are absolutely correct. He is grinning and smiling and laughing, experiencing intense pleasure. He does not want it to stop. He rolls onto his back and waves his four paws in the air. The children scream with laughter, so he does it even more enthusiastically. His grin grows larger, his smile broader, his laughter louder. He is in ecstasy.

Anthropomorphism? Call it what you will, but Benjy is experiencing pure pleasure. He loves the attention. He loves the children to make a fuss over him. *They* have no problem reading his face. It may not be a human grin, a human smile, or human laughter, but the facial gestures Benjy makes are crystal clear to these youngsters: The dog is *happy*. He loves the attention. The dog licks them and urges them on. The dog wants to get pleasure and he also wants to give it. The children are happy to give the dog pleasure and to take pleasure from the dog. They love the dog. The dog loves them.

There are writers who consider a dog's love for humans as an anthropomorphic conceit, but human love for dogs to be not only true but for some, the deepest love they will ever experience. Earlier I quoted George Steiner: "What can be absolute is our love for the animal or animals in our lives, asking for no guaranteed return." I am not sure there are many humans who can love without expecting a return. But surely it is in the very nature of dogs to do precisely that: They love us with the hope—but no guarantee—that we will love them back and treat them well.

Love. It is the most elusive word in the human vocabulary. We cannot get by without it, even if we can't define it. We may not always be able to recognize it. We have trouble talking about it. All of us, just about, have the sense that the word defines our species. That it makes everything else less important. In the end, life comes down to love. Not how much money we have. Not how much social standing we have. Not even how healthy we are. Not how many possessions, or even how many children. Only: Have we loved? Are we loved? Do we have love now? Perhaps the most common words of a person dying are "I love you." Certainly the most common words said to a dying person are "I love you."

Do we see and acknowledge love in other species? We see an entire elephant herd wait for the young elephant who is crippled and cannot quite keep up. We see the young elephant smell the bones of her mother and linger over them. We *know* for sure that she is experiencing something like what we call love.

But for some reason it is harder than it should be for humans to acknowledge love in other species. Elephants are an exception.

The writing of some great elephant researchers has been very persuasive and some of the footage many of us have seen is irrefutable. But if you try to make the same claims for cows and their calves, ewes and their lambs, or just about any other animal, you will encounter a chorus of skeptical voices. And if you attempt to speak about love that crosses the species barrier, well, just try that at the next meeting of the Animal Behavioral Society and see what kind of response you get.

There is one great exception: dogs. Exactly why it is that people are less likely to deny love in dogs is still something of a mystery. It remains true that you probably won't fare very well at that same animal behavior meeting talking about love or empathy or compassion in dogs. There will be cries of anthropomorphism. But there will be fewer of them than there were twenty years ago. True, there will be some scientists who will ridicule the idea of love in any nonhuman animal, but they will no longer be in the majority. Cynologists (dog experts) might trot out a golden cliché that has proved its value over many years and in many cultures: Dogs are *programmed* to feel love. They come from a highly sociable forebear, the gray wolf, and are the descendants of hierarchically obsessed animals: It is in their interest to appease a more powerful packmate, to beg for attention and affection in the form of kisses and moans and groans of pleasure. It is but a trick, a deception, nothing but a hardwired attempt at pacification and mollification. It is pure suck.

But they're wrong. Perhaps they are right when it comes to wolves—I haven't lived with wolves, so I wouldn't know. Nobody has ever lived with a free-roaming wolf pack, so our scientific knowledge of wolf love is pretty limited. But I have lived with dogs, many of them, and I live with one particular dog right now who to me is pure love.

When Benjy licks our hands and our faces and attempts to kiss our mouths, he is not soliciting food or appeasement; he is soliciting and expressing love. I don't know if he thinks about love, if he contemplates it the way we do, if he searches his memories for it. I don't even know if he knows that it sustains him. But he knows that we feel it for him and that he feels it for us. Or maybe he doesn't *know* it in any recognizable sense of the word, but the love is no less true for that. *We* know it for sure. Every day he expresses it. Every gesture proclaims it. You know what I am talking about, those of you who live with dogs. It is the rare dog who does not express love. I am not talking about affection, or friendship, or play, though love can be expressed through all of these. I am talking about that most profound of all human concepts, love.

But if I am right, there is still the lingering question of why, of all animals, only dogs should have the ability to display love in much the same way as humans. That is the theme of this book: They do it because they have been our constant companions for the last 40,000 years, and during that long romance we reinforced one another's finest qualities. Dogs can be vicious, but we love best the dogs who are not; humans can be cruel, but dogs love best the ones who are not. Over those years we tended to extend our love beyond the boundaries of our own tribe—indeed, beyond the boundaries of our species. It is an enormous leap. I would even say it is one of the great evolutionary saltations of all time, a quantum step beyond where we or any other species had ever managed to go. I don't believe we got there entirely on our own. As a species, we are a bit like the great King Yudhishtira, of whom I spoke in an earlier chapter. Just as he was accompanied on his last journey by a simple unnamed dog, so we have been accompanied by dogs along the full trajectory of becoming human. In fact, it is not just that they accompanied us: Sometimes they led us and sometimes we led them. But we made the trip together. And that unique voyage changed both of us: We became the two species that have nearly merged. We are dogmen. Dogs are our totems. They may regard us as more than human. But any godly characteristics we may possess, we possess because of them. (OK. And mothers!)

POSTSCRIPT

A WALK ON THE BEACH WITH BENJY AND THE CATS

I finished typing the last lines of this book on a quiet, warm spring night around midnight. I had been thinking about the topic and writing for a long time, so it was a relief to finally finish. The waves were quietly lapping at the shore of the beach in front of our house and Benjy was waiting for his late-night walk. As I headed out the door with him, the three cats followed. The moon was full and the beach was flooded with light. The five of us were alone on the beach. The cats, for some reason, were especially frisky and kept hiding behind bushes on the sand and ambushing Benjy. He knew it was all in good fun and smiled with his characteristic good humor.

Suddenly I felt overcome with happiness. I realized just how much I loved these late-night walks on the beach with my four friends. I realized too just how much pleasure they took from the walks as well. The animlas came completely alive. There was a freedom in everyone's behavior not as noticeable at other times. There was nobody else around, nothing to worry about and all four animals (five, actually) lost their reserve. The three cats did something they only do on the beach and even then only occasionally: They galloped at full speed toward Benjy and me and then toward one another for the sheer fun of it. It was contagious. Benjy grabbed a long stick (log was more like it) and pranced around with it. I sang (off-key) about how much I adored them all.

I was in a state of bliss. I was observing and was part of the true companionship that can come about when three species are completely comfortable with one another. Difference was forgotten for this half hour. Five neotenic animals from three neotenic species playing happily on a beach, bound by one thing only: love. We all felt it. I knew it would have to pass eventually, but for that moment I knew the purpose of life: to love as many other sentient beings as possible and to feel the love reciprocated. It was as if the secret of the book I was writing was suddenly revealed to me and embodied on this single moonlit walk on the beach, bound by love.

ACKNOWLEDGMENTS

The following people have been particularly helpful: In first place comes an extraordinary agent, Andy Ross, never too busy to respond immediately to any request for help, guidance, opinions, or opera tickets! And who knows books better than the former owner of America's late but best-loved bookstore, Cody's Books? Professor Wolfgang Schleidt, the long-time assistant of the legendary zoologist and Nobel Prize winner Konrad Lorenz, who holds a theory very similar to mine and who has inspired many generations of researchers, was quick to give me leads and express his views. Professor Anatoly Ruvinsky provided a lovely quote convincing me I was on the right track. Marc Bekoff sent me many of his articles, all seminal for understanding dog psychology, as did Sunil Kumar Pal. Juliet Clutton-Brock, the foremost expert on domestication, encouraged my research and found it plausible. Professors Lesley Rogers and Gisela Kaplan took an afternoon off to discuss these issues with me in the presence of their four enchanting dogs and gave me a copy of their indispensable *Wild Dogs*.

Jessica Walker and Dale Arnja of UNITEC New Zealand found me many an obscure article when everyone else had given up, and talked with me about many of the main themes of the book. Janice Koler-Matznick, the leading authority on the New Guinea singing dog and a dog behaviorist, commented on an early draft of the book, as did Adam Miklosi of Budapest University. Robert Wayne, the world's leading authority on dog genetics, gave me his latest thinking on the dating of domestication. My old friend, professor Robert Goldman, an authority on ancient India, gave me his impressions of pariah dogs in Delhi. I am grateful to the Royal New Zealand Foundation of the Blind for allowing us to "re-home" Benjy and for providing him with early training. I am particularly indebted to Belinda Simpson. Not only did she introduce our family to Benjy, but she talked to me about seizure-alert dogs, trained Benjy to be a therapy dog, has offered him comfort, affection, and understanding, and is herself deeply in tune with canine love. I wish to also thank my friend Cornelia Dodssuweit, a dog behaviorist, for many pleasant and fruitful discussions about Benjy and other dogs. My old friend Marti Kheel read the book in manuscript and gave me many valuable suggestions for improvement. The American writer Richard de-Grandpre (who now lives in Auckland) helped me at a crucial time in the preparation of the manuscript by reading and advising. Many thanks for his many skills.

I would also like to thank my editor, Allison Lorentzen, for her light and sure touch, and all the other terrific people at HarperCollins, Jonathan Burnham, Kathy Schneider, Tina Andreadis, Angie Lee, Nicole Reardon, Amy Vreeland (who did such a great job of line-editing), Tom McNellis, and Christine Van Bree, for all they have done to make this book a success and remind people of how lucky we are to have dogs in our lives!

My daughter, Simone, works with animals every day, and has much to teach me. My younger sons, Ilan and Manu, produce illuminating insights by constantly questioning me. My wife, Leila, remains my soul mate. Her only rival is Benjy, the dog who could not, would not, has not stopped loving. It seems only fair to dedicate this book to him.

APPENDIX

BENJY'S OFFICIAL ASSESSMENT BY THE ROYAL NEW ZEALAND FOUNDATION OF THE BLIND, GUIDE DOG SERVICES

Benjy was born at the Royal New Zealand Foundation of the Blind, Guide Dog Services, Guide Dog Centre on 1 October 2005. His father was a sire called Normandy (imported from a U.S. guide dog school) and his mother was a dam called Josie (a proven GDS breeding dog). Benjy remained in the breeding center with his mother and littermates until about 8 weeks of age and then he was placed in a puppy walking home in Auckland.

From December 2005 to November 2006 Benjy was placed with a volunteer puppy-walking family in Auckland. During this time he was taken to a variety of places to allow exposure to the wide range of sights and sounds that he would encounter as a working guide dog.

Benjy and his puppy walker received monthly visits from a GDS puppy development supervisor and they also attended puppy classes that were designed to develop and enhance his dog interaction and obedience skills.

During the puppy classes he generally would mix well with the other dogs and appeared to enjoy this socialization. However, it was noted that he showed some lethargy during these sessions and was not overly interactive with other dogs, and where the opportunity arose he would scan the ground and look for food.

In the early stages of his development Benjy was like any other puppy and initially appeared confident in the environment and fairly laid back. However, over time he began to present as a more sensitive type of dog with above average levels of anxiety. He also showed reluctance toward having his guide dog puppy coat placed on his back. All of this impacted on his level of confidence.

His level of motivation was variable, depending on the activity he was involved in and the time of day. He would often stop on walks and refuse to lead out, indicating he didn't really want to walk at that time or at that place; he would happily lead out if going to a park for a free run.

Over time all these reactions began to improve and it was hoped that once he entered training he would enjoy the stimulus of finding destinations and the responsibility associated with being a working guide dog.

Benjy arrived at the GDC to begin his initial assessment and in-depth training program on November 16, 2006. During this time he was taken to a variety of areas to assess how he reacted and coped. These included residential, city, buses, trains, lifts, stairs, escalators, shopping malls, and areas where there are other dogs.

He was evaluated on a variety of temperament characteristics but unfortunately did not score well on assessments relating to anxiety, body sensitivity, and motivation, and as a result he was considered unsuitable for guiding work. Benjy was a lovely dog to be around, but his lackadaisical approach was better suited to a more laid-back career. On November 22, 2006 he was officially withdrawn from the program and transferred to the GDS Adoption Programme.

After a period of time boarding locally, Benjy was offered to NZ Epilepsy Assistance Dogs Trust for training. Unfortunately this was not successful and Benjy was returned to Guide Dog Services and placed for adoption. On September 8, 2007 he was adopted by Dr. Jeffrey Masson and placed in a home where he gets to spend quality time doing the things he loves to do—playing, walking on the beach, and being around people.

BIBLIOGRAPHY

These are the primary books I have used in my research on dogs. For those not obsessed with books, here are the books I find most useful, followed by the rest.

Coppinger, Raymond, and Lorna Coppinger. *Dogs: A Startling New Understanding of Canine Origin, Behavior, and Evolution*. New York: Scribner, 2001.

Csányi, Vilmos. *If Dogs Could Talk: Exploring the Canine Mind*. New York: Farrar, Straus & Giroux, 2005.

Derr, Mark. *Dog's Best Friend: Annals of the Dog-Human Relationship*. Chicago: University of Chicago Press, 2004.

Déry, Tibor. *Nikki, the Story of a Dog*. New York Review of Books Classics, 2009. First published in Hungarian in 1956.

Fiennes, Richard, and Alice Fiennes. *The Natural History of Dogs*. Garden City, NY: Natural History Press, 1970.

Fox, Michael W. *Behavior of Wolves, Dogs and Related Canids*. New York: Harper & Row, 1971.

Franklin, Jon. *The Wolf in the Parlor: The Eternal Connection Between Humans and Dogs*. New York: Henry Holt & Co., 2009.

Horowitz, Alexandra. *Inside of a Dog: What Dogs See, Smell, and Know*. New York: Scribner, 2009.

Kerasote, Ted. *Merle's Door: Lessons from a Freethinking Dog*. New York: Harcourt, Inc., 2007.

Lorenz, Konrad. *Man Meets Dog.* With a new introduction by Donald McCaig; translated by Marjorie Kerr Wilson. New York: Kodansha America, 1994.

Miklosi, Adam. *Dog Behaviour, Evolution, and Cognition.* Oxford, UK: Oxford University Press, 2007.

Page, Jack. *Dogs: A Natural History.* Washington, DC: Smithsonian Books, 2007.

Scott, J. P., and J. L. Fuller. *Genetics and the Social Behavior of the Dog.* Chicago: University of Chicago Press, 1965.

Quinn, Spencer. *Dog on It.* New York: Atria Books, 2009.

Thomas, Elizabeth Marshall. *The Hidden Life of Dogs.* Boston: Houghton Mifflin Co., 1993.

Thurston, Mary Elizabeth. *The Lost History of the Canine Race: Our 15,000-Year Love Affair with Dogs.* Kansas City, MO: Andrews & McMeel, 1996.

STARRED (*) ITEMS HAVE BEEN PARTICULARLY USEFUL.

Abrantes, Roger. *The Evolution of Canine Social Behavior, 2nd ed.* Ann Arbor, MI: Wakan Tanka Publishers, 2005.

*Anderson, P. Elizabeth. *The Powerful Bond between People and Pets: Our Boundless Connections to Companion Animals.* Westport, CT: Praeger, 2008.

Arkow, Phil, ed. *The Loving Bond: Companion Animals in the Helping Professions.* Saratoga, CA: R & E Publishers, 1987.

Arluke, Arnold, and Clinton R. Sanders. *Regarding Animals.* Philadelphia: Temple University Press, 1996.

*von Arnin, Elizabeth. *All the Dogs of My Life.* London: Virago Press, 1995.

Auster, Paul. *Timbuktu: A Novel.* New York: Henry Holt & Co., 1999.

Bauer, Nona Kilgore. *Dog Heroes of September 11th: A Tribute to America's Search and Rescue Dogs.* Allenhurst, NJ: Kennel Club Books, 2006.

Beck, Alan, and Aaron Katcher. *Between Pets and People: The Importance of Animal Companionship.* Revised ed., with a foreword by Elizabeth Marshall Thomas. West Lafayette, IN: Purdue University Press, 1996.

Benjamin, Carole Lea. *Mother Knows Best: The Natural Way to Train Your Dog.* New York: Howell Book House, 1985.

Berger, John. *King: A Street Story.* New York: Pantheon, 1999.

Bergler, Reinhold. *Man and Dog: The Psychology of a Relationship* (translation of *Mensch und Hund*). New York: Howell Book House, 1989.

Boitani, Luigi. *De la parte de lupo*. Milan: Giorgio Mondadori, 1986.

Bollée, Willem. *Gone to the Dogs in Ancient India*. Munich: C. H. Beck, 2006.

Budiansky, Stephen. *The Truth about Dogs: An Inquiry into the Ancestry, Social Conventions, Mental Habits, and Moral Fiber of Canis familiaris*. New York: Penguin, 2000.

Bueler, Lois E. *Wild Dogs of the World*. New York: Stein & Day, 1973.

Bush, Karen. *Everything Dogs Expect You to Know*. Sydney: New Holland, 2007.

Buytendijk, Frederik. *The Mind of the Dog*. Translated by Lillian A. Clare. Boston: Houghton Mifflin Co., 1936.

*Caras, Roger. *A Dog Is Listening: The Way Some of Our Closest Friends View Us*. New York: Summit Books, 1992.

Choron, Sandra, and Harry Choron. *Planet Dog: A Doglopedia*. Boston: Houghton Mifflin Co., 2005.

Clutton-Brock, Juliet, and Kim Dennis-Bryan. *Dogs of the Last Hundred Years at the British Museum (Natural History)*. London: British Museum, 1988.

Coppinger, Raymond. *Fishing Dogs*. London: Secker & Warburg, 1997.

*Coppinger, Raymond, and Lorna Coppinger. *Dogs: A Startling New Understanding of Canine Origin, Behavior, and Evolution*. New York: Scribner, 2001.

*Corbett, Laurie. *The Dingo in Australia and Asia*. Marleston, South Australia: J. B. Books, 2001.

Coren, Stanley. *The Modern Dog: How Dogs Have Changed People and Society and Improved Our Lives*. New York: Free Press, 2008.

———. *Why We Love the Dogs We Do: How to Find the Dog that Matches Your Personality*. New York: Simon & Schuster, 1998.

*Crisler, Lois. *Arctic Wild*. New York: Harper Brothers, 1958.

*Csányi, Vilmos. *If Dogs Could Talk: Exploring the Canine Mind*. New York: Farrar, Straus & Giroux, 2005.

Cusack, Odean. *Pets and Mental Health*. New York: Haworth Press, 1988.

Debroy, Bibek. *Sarama and Her Children: The Dog in Indian Myth*. New Delhi: Penguin, 2008.

Delise, Karen. *Fatal Dog Attacks: The Stories Behind the Statistics*. Manorville, NY: Anubis Press, 2002.

———. *The Pit Bull Placebo: The Media, Myths and Politics of Canine Aggression*. Manorville, NY: Anubis Press, 2007.

*Derr, Mark. *Dog's Best Friend: Annals of the Dog-Human Relationship*. Chicago: University of Chicago Press, 2004.

———. *A Dog's History of America: How Our Best Friend Explored, Conquered, and Settled a Continent*. New York: North Point Press, 2004.

*Déry, Tibor. *Nikki, the Story of a Dog*. New York Review of Books Classics, 2009. First published in Hungarian in 1956.

Dodman, Nicholas. *The Dog Who Loved Too Much: Tales, Treatments, and the Psychology of the Dog*. New York: Bantam, 1996.

*Doty, Mark. *Dog Years: A Memoir*. New York: HarperCollins, 2007.

Duemer, Joseph, and Jim Simmerman, eds. *Dog Music: Poetry about Dogs*. New York: St. Martin's Press, 1996.

Dunbar, Ian, and Michael Berman. *Dog Behavior: Why Dogs Do What They Do*. Neptune, NJ: T. F. H. Publications, 1979.

Dye, Dan, and Mark Beckloff. *Amazing Gracie: A Dog's Tale*. New York: Workman, 2000.

Editors of The Bark. *Dog Is my Co-Pilot: Great Writers on the World's Oldest Friendship*. New York: Three Rivers Press, 2003.

*Feddersen-Petersen, Dorit. *Ausdrucksverhalten beim Hund*. Jena: Gustav Fischer, 1995.

*Fiennes, Richard, and Alice Fiennes. *The Natural History of Dogs*. New York: Bonanza Books, 1968.

*Fischer, Werner. *Die Seele des Hundes*. Berlin-Hamburg: Paul Parey, 1961.

Fogle, Bruce. *The Dog's Mind: Understanding Your Dog's Behavior*. New York: Howell Book House, 1990.

*Fox, Michael W. *Behavior of Wolves, Dogs and Related Canids*. New York: Harper & Row, 1971.

———. *Dog Body, Dog Mind: Exploring Your Dog's Consciousness and Total Well-Being*. Guilford, CT: Lyons Press, 2007.

*Frank, H., ed. *Man and Wolf: Advances, Issues, and Problems in Captive Wolf Research*. Dordrecht, Netherlands: W. Jung Publishers (Kluwer Academic Publishers Group), 1987.

*Franklin, Jon. *The Wolf in the Parlor: The Eternal Connection Between Humans and Dogs*. New York: Henry Holt & Co., 2009.

Fudge, Erica. *Animal*. London: Reaktion Books, 2002.

Gaita, Raimond. *The Philosopher's Dog: Friendships with Animals*. New York: Random House, 2002.

*Garber, Marjorie. *Dog Love*. New York: Simon & Schuster, 1996.

Gary, Romain. *White Dog*. London: Jonathan Cape, 1971.

Genoways, Hugh, and Marion Burgwin, eds. *Natural History of the Dog*. Pittsburgh, PA: Carnegie Museum of Natural History, 1984.

Gentry, Christine. *When Dogs Run Wild: The Sociology of Feral Dogs and Wildlife*. Jefferson, NC: McFarland & Co., 1983.

Gibbs, Margaret. *Leader Dogs for the Blind*. Fairfax, VA: Denlinger's Publishers, 1982.

*Gonzales, Philip, and Leonore Fleischer. *The Dog Who Rescues Cats: The True Story of Ginny*. Introduction by Cleveland Amory. New York: HarperCollins, 1995.

Gray, John. *Straw Dogs: Thoughts on Humans and Other Animals*. London: Granta Books, 2002.

Grier, Katherine C. *Pets in America: A History*. Chapel Hill, NC: University of North Carolina Press, 2006.

Grogan, John. *Bad Dogs Have More Fun: Selected Writings on Family, Animals, and Life from The Philadelphia Inquirer*. New York: Perseus Press, 2007.

*———. *Marley & Me: Life and Love with the World's Worst Dog*. New York: William Morrow, 2005.

Grossman, Loyd. *The Dog's Tale: A History of Man's Best Friend*. London: BBC Books, 1993.

Hales, Steven D., ed. *What Philosophy Can Tell You About Your Dog*. Chicago: Open Court, 2008.

Hall, Roberta L., and Henry S. Sharp, eds. *Wolf and Man: Evolution in Parallel*. New York: Academic Press, 1978.

Hampton, Bruce. *The Great American Wolf*. New York: Henry Holt & Co., 1997.

Haraway, Donna. *The Companion Species Manifesto: Dogs, People, and Significant Otherness*. Chicago: University of Chicago Press (Prickly Paradigm Press), 2003.

———. *When Species Meet*. Minneapolis: University of Minnesota Press, 2008.

Hearne, Vicki. *Bandit: Dossier of a Dangerous Dog*. New York: HarperCollins, 1991.

———. *The White German Shepherd*. New York: Atlantic Monthly Press, 1988.

Hearst, Dorothy. *Promise of the Wolves: The Wolf Chronicles*. New York: Simon & Schuster, 2008.

*Horowitz, Alexandra. *Inside of a Dog: What Dogs See, Smell, and Know*. New York: Scribner, 2009.

*Humphrey, E. S., and L. Warner. *Working Dogs*. Baltimore, MD: Johns Hopkins Press, 1934.

Irvine, Leslie. *If You Tame Me: Understanding Our Connection with Animals*. Philadelphia: Temple University Press, 2004.

Jensen, Per, ed. *The Behavioural Biology of Dogs*. Oxfordshire, UK: Cabi Publishing, 2007.

Katz, Jon. *Dog Days: Dispatches from Bedlam Farm*. New York: Villard, 2007.

———. *A Good Dog: The Story of Orson, Who Changed My Life*. New York: Random House, 2007.

———. *Izzy & Lenore: Two Dogs, an Unexpected Journey, and Me*. New York: Villard, 2008.

Karlekar, Hiranmay. *Save Humans and Stray Dogs: A Study in Aggression*. Los Angeles: Sage, 2008.

*Katcher, Aaron Honori, and Alan M. Beck, eds. *New Perspectives on Our Lives with Companion Animals*. Philadelphia: University of Pennsylvania Press, 1983.

Klinghammer, Erich, ed. *The Behavior and Ecology of Wolves*. New York: Garland STPM, 1979.

*Kerasote, Ted. *Merle's Door: Lessons from a Freethinking Dog*. New York: Harcourt, Inc, 2007.

*Knapp, Caroline. *Pack of Two: The Intricate Bond Between People and Dogs*. New York: Delta, 1999.

Kuusisto, Stephen. *Planet of the Blind: A Memoir*. New York: Dell, 1998.

Kuzniar, Alice A. *Melancholia's Dog: Reflections on our Animal Kinship*. Chicago: University of Chicago Press, 2006.

*Leach, Maria. *God Had a Dog: Folklore of the Dog*. New Brunswick, NJ: Rutgers University Press, 1961.

Lemish, Michael G. *War Dogs: Canines in Combat*. McLean, VA: Brassey's (division of Macmillan), 1996.

*Lerman, Rhoda. *In the Company of Newfies: A Shared Life*. New York: Henry Holt & Co., 1996.

*Lindsay, S. R. *Handbook of Applied Dog Behavior and Training* (3 vols.). Ames, IA: Blackwell Publishing, 2000–2005.

Lopez, Barry. *Of Wolves and Men*. New York: Charles Scribner's Sons, 1987.

*Lorenz, Konrad. *Man Meets Dog*. With a new introduction by Donald McCaig. Translated by Marjorie Kerr Wilson. New York: Kodansha America, 1994.

Mann, Thomas. *A Man and His Dog*. New York: Alfred A. Knopf, 1930. (Originally published as *Herr und Hund* by S. Fischer, 1918, translated by Herman George Scheffauer.)

Markoe, Merrill. *Walking in Circles Before Lying Down: A Novel*. New York: Villard, 2006.

———. *What the Dogs Have Taught Me*. New York: Viking, 1992.

Mayle, Peter. *A Dog's Life*. New York: Knopf, 1995.

McCaig, Donald. *Nop's Trials.* New York: Crown Publishers, 1984.

McConnell, Patricia B. *For the Love of a Dog: Understanding Emotion in You and Your Best Friend.* New York: Ballantine, 2006.

———. *The Other End of the Leash: Why We Do What We Do Around Dogs.* New York: Ballantine, 2002.

McEwan, Ian. *Black Dogs.* New York: Doubleday, 1992.

*McHugh, Susan. *Dog.* London: Reaktion Books, 2004.

McLoughlin, John C. *The Canine Clan: A New Look at Man's Best Friend.* New York: Viking, 1983.

Mech, L. David. *The Wolf: The Ecology and Behavior of an Endangered Species.* Minneapolis: University of Minnesota Press, 1981.

*Mech, L. David, and Luigi Boitani, eds. *Wolves: Behavior, Ecology, and Conservation.* Chicago: University of Chicago Press, 2003.

Merritt, Raymond, and Miles Barth. *A Thousand Dogs.* Cologne: Taschen, 2008.

Michalko, Rod. *The Two in One: Walking with Smokie, Walking with Blindness.* Philadelphia: University of Pennsylvania Press, 1999.

*Miklosi, Adam. *Dog Behaviour, Evolution, and Cognition.* Oxford, UK: Oxford University Press, 2007.

Miles, Kathryn. *Adventures with Ari: A Puppy, a Leash and Our Year Outdoors.* New York: Skyhorse Publishing, 2009.

Millan, Cesar (with Melissa Jo Peltier). *Be the Pack Leader.* New York: Crown, 2007.

*Morris, Desmond. *Dogwatching.* New York: Crown, 1986.

Neville, Peter. *Do Dogs Need Shrinks?* New York: Carol Publishing Group, 1992.

*Newby, Jonica. *The Pact for Survival: Humans and Their Animal Companions.* Sydney: ABC Books, 1997.

*Olmert, Meg Daley. *Made for Each Other: The Biology of the Human-Animal Bond.* New York: Perseus, 2009.

Olsen, Stanley J. *Origins of the Domestic Dog: The Fossil Record.* Tucson, AZ: University of Arizona Press, 1985.

Ostrander, Elaine A., Urs Giger, Kerstin Lindblad-Toh, eds. *The Dog and its Genome.* Cold Spring Harbor, NY: Cold Spring Harbor Laboratory Press, 2006.

*Page, Jack. *Dogs: A Natural History.* Washington, DC: Smithsonian Books, 2007.

*Pfaffenberger, C. J. *The New Knowledge of Dog Behavior.* New York: Howell Books, 1963.

*Podberscek, Anthony L., Elizabeth S. Paul, James A. Serpell, eds. *Com-*

panion Animals and Us: Exploring the Relationships between People and Pets. Cambridge: Cambridge University Press, 2000.

Porter, V. *Faithful Companions: The Alliance of Man and Dog*. London: Methuen, 1989.

Putnam, Peter Brock. *Love in the Lead: The Miracle of the Seeing Eye Dog*. New York: University Press of America, 1997.

Quindlen, Anna. *Good Dog. Stay*. New York: Random House, 2007.

*Quinn, Spencer. *Dog on It*. New York: Atria Books, 2009.

Rivera, Michelle A. *Canines in the Classroom: Raising Humane Children through Interactions with Animals*. New York: Lantern Books, 2004.

Robinson, I., ed. *The Waltham Book of Human-Animal Interaction: Benefits and Responsibilities of Pet Ownership*. Oxford, UK: Pergamon, 1995.

*Rogers, Lesley J., and Gisela Kaplan. *Spirit of the Wild Dog: The World of Wolves, Coyotes, Foxes, Jackals and Dingoes*. London: Allen & Unwin, 2003.

Rosen, Michael J., ed. *The Company of Dogs: 21 Stories by Contemporary Masters*. New York: Doubleday, 1990.

———. *Dogs We Love*. New York: Artisan, 2008.

Rowlands, Mark. *The Philosopher and the Wolf: Lessons from the Wild on Love, Death, and Happiness*. London: Granta, 2008.

Sanders, Clinton R. *Understanding Dogs: Living and Working with Canine Companions*. Philadelphia: Temple University Press, 1999.

Sax, Boria. *Animals in the Third Reich: Pets, Scapegoats, and the Holocaust*. New York: Continuum, 2000.

Schaffer, Michael. *One Nation Under Dog: Adventures in the New World of Prozac-Popping Puppies, Dog-Park Politics, and Organic Pet Food*. New York: Henry Holt & Co., 2009.

Schwartz, Marion. *A History of Dogs in the Early Americas*. New Haven, CT: Yale University Press, 1997.

*Scott, J. P., and J. L. Fuller. *Genetics and the Social Behavior of the Dog*. Chicago: University of Chicago Press, 1965.

Sheehan, Jacqueline. *Lost & Found*. New York: HarperCollins, 2007.

*Sheldon, Jennifer W. *Wild Dogs: The Natural History of the Nondomestic Canidae*. Caldwell, NJ: Blackburn Press, 1992.

Sherrill, Martha. *Dog Man: An Uncommon Life on a Faraway Mountain*. New York: Penguin, 2008.

Shojai, Amy. *A Dog's Life: The History, Culture, and Everyday Life of the Dog*. New York: Friedman/Fairfax, 1994.

*Smythe, R. H. *The Mind of the Dog*. Springfield, IL: Charles C. Thomas, 1961.

Stafford, Kevin. *The Welfare of Dogs*. Dordrecht, Netherlands: Springer, 2007.

Stein, Garth. *The Art of Racing in the Rain: A Novel*. New York: HarperCollins, 2008.

*Steinhart, Peter. *In the Company of Wolves*. New York: Knopf, 1995.

Summers, Judith. *My Life with George: What I Learned about Joy from One Neurotic (and Very Expensive) Dog*. New York: Hyperion, 2007.

*Thomas, Elizabeth Marshall. *The Hidden Life of Dogs*. Boston: Houghton Mifflin Co., 1993.

*Thurston, Mary Elizabeth. *The Lost History of the Canine Race: Our 15,000-Year Love Affair with Dogs*. Kansas City, MO: Andrews & McMeel, 1996.

*Trumler, Eberhard. *Understanding Your Dog*. Translated from the German by Richard Barry. London: Faber & Faber, 1973.

*———. *Your Dog and You*. New York: Seabury, 1973.

*Vesey-Fitzgerald, Brian, ed. *The Book of the Dog*. Toronto: Borden Publishing, 1948.

*Wang, Xiaoming, and Richard H. Tedford. *Dogs: Their Fossil Relatives and Evolutionary History*. New York: Columbia University Press, 2008.

Watson, Brad. *Last Days of the Dog-Men*. London: Weidenfeld & Nicolson, 1997.

Willis, Roy, ed. *Signifying Animals: Human Meaning in the Natural World*. London: Routledge, 1990.

Wilson, Cindy C., and Dennis C. Turner, eds. *Companion Animals in Human Health*. London: Sage, 1998.

*Winograd, Nathan J. *Redemption: The Myth of Pet Overpopulation and the No Kill Revolution in America*. Los Angeles: Almaden, 2007.

*Winokur, Jon., ed. *Mondo Canine*. New York: E. P. Dutton, 1991.

Woloy, Eleanora M. *The Symbol of the Dog in the Human Psyche*. Wilmette, IL: Chiron Publications, 1990.

Wroblewski, David. *The Story of Edgar Sawtelle: A Novel*. New York: HarperCollins, 2008.

*Zimen, Erik. *Der Hund: Abstammung—Verhalten—Mensch und Hund*. Munich: C. Bertelsmann, 1988.

*———. *The Wolf: His Place in the Natural World*. Translated from the German by Eric Mosbacher. London: Souvenir Press, 1981.

NOTES

xi The Harvard Biologist E. O. Wilson: *The Biophilia Hypothesis*, edited by Stephen R. Kellert and Edward O. Wilson. Washington, DC: Island Press/Shearwater Books, 1993.

xvi He marked a passage: Carl Sagan, in his book *Shadows of Forgotten Ancestors*, New York: Random House, 1992, p. 371, cites this same passage and goes on to make the memorable statement that "Humans—who enslave, castrate, experiment on, and fillet other animals—have had an understandable penchant for pretending that animals do not feel pain."

16 Christian greeted them: *A Lion Called Christian: The True Story of the Remarkable Bond Between Two Friends and a Lion*, by Anthony Bourke, John Rendall, with a foreword by George Adamson. New York: Broadway Books, 2009. The book was first published in 1971. Amazon has the video on its site where the book is advertised.

16 Tigers and lions: For a list from Zoocheck, a respected organization that monitors such matters (their motto: "Life in the circus is no life at all"). See www.zoocheck.com/campaigns_circuses_incidents%2097-10.html.

16 Here was an animal born and raised: For an excellent discussion of why these animals should never be kept in captivity, let alone considered "pets," see the Humane Society of the United States' statement: www.hsus.org/wildlife/wildlife_news/siegfried_roy_incident_underscores_the_dangers_of_exotic_pets.html.

16 It is not only the cat family: A wonderful book (don't be misled by the pedestrian title) is Stephen Herrero's *Bear Attacks: Their Causes and Avoidance*. Guilford, CT: Lyons Press, 2002. Written by the world's leading authority on bears, it is actually an academic and scientific treatise on the animals and makes you realize just how unpredictable they can be. If you plan to ever be around bears in the wild, you need to read this book first.

17 In 2003: I write about this in my book *Altruistic Armadillos, Zenlike Zebras*. New York: Ballantine, 2006. See Timothy Treadwell (with Jewel Pavolak), *Among Grizzlies: Living with Wild Bears in Alaska*. New York: Ballantine, 1999.

17 Possibly because we have destroyed: See Charles Siebert, "An Elephant Crackup?" *New York Times*, October 8, 2006.

25 Nikki spends most of his time: Tibor Déry, *Nikki, the Story of a Dog*. Introduction by George Szirtes; translated by Edward Hyams. New York: New York Review of Books, 2009, p. 80.

26 When it comes to play: See the elaborate explanation and description of dog play in Alexandra Horowitz's book *Inside of a Dog: What Dogs See, Smell, and Know*. New York: Scribner, 2009.

27 I have given this a great deal of thought: We can find much evidence for such sensitivity in the new book by Alexandra Horowitz, *Inside of a Dog: What Dogs See, Smell, and Know*. New York: Scribner, 2009.

34 There is that touching moment: By an extraordinary coincidence the same thing happened to Sigourney Weaver during the actual filming with wild gorillas. The large silverback reached out and touched Weaver's hand. Michael Apted, the director of the film, told me he believed the gorilla remembered Dian Fossey, and thought she had returned!

38 People in India: See Bibek Debroy, *Sarama and Her Children: The Dog in Indian Myth*. New Delhi: Penguin, 2008. Also the remarkable scholarly book by Willem Bollée, *Gone to the Dogs in Ancient India*. Munich: C. H. Beck, 2006.

42 before other animals: Susan Janet Crockford, a zoologist specializing in the domestication of the dog, in a paper delivered to the 30th World Congress of the World Small Animal Veterinary Association in 2005, speculates that "dogs appear to have been generated on at least three separate occasions (perhaps more) from geographically distinct ancestral popula-

tions of wolf (i.e., different subspecies)." Her paper can be retrieved at www.vin.com/proceedings/Proceedings.plx?CID=WSAVA2005&PID=11071&O=Generic.

43 In the past fifteen years: C. Vilà, R. K. Wayne, et al., "Multiple and ancient origins of the domestic dog." *Science*, 276: 1687–89, 1997.

43 anywhere from 15,000 years to 150,000: Here is a summary from the work of Dr. Peter Savolainen from the School of Biotechnology at the Royal Institute of Technology (KTH) in Stockholm (www.biotech.kth.se/genetech/groups/savolainen_group.html#Publications):

"Several 10–15,000 year old archaeological finds in Eastern Europe and the Middle East of domestic dog-looking canids indicate that the dog was the first domestic animal. Apart from this, however, archaeological studies have not produced any detailed facts about the origin of the domestic dog. Such basic questions as the number of founding events, and where and when these occurred, have remained unsolved. To address these questions we initiated a cooperation study of mtDNA sequence variation among domestic dogs, based on our comprehensive collection of more than 1,000 DNA samples from dogs from most parts of the world. We could thereby show that the geographic variation of mtDNA among domestic dogs indicates that the dog originates from East Asia, probably about 15,000 years ago." See "Genetic evidence for an East Asian origin of domestic dogs." Peter Savolainen, et al., *Science*, November 22, 2002, vol. 298, no. 5598, pp. 1610–13. The summary reads as follows: "The origin of the domestic dog from wolves has been established, but the number of founding events, as well as where and when these occurred, is not known. To address these questions, we examined the mitochondrial DNA (mtDNA) sequence variation among 654 domestic dogs representing all major dog populations worldwide. Although our data indicate several maternal origins from wolf, >95% of all sequences belonged to three phylogenetic groups universally represented at similar frequencies, suggesting a common origin from a single gene pool for all dog populations. A larger genetic variation in East Asia than in other regions and the pattern of phylogeographic variation suggest an East Asian origin for the domestic dog, ~15,000 years ago." Note this: "Dr. Peter Savolainen, a former colleague of Dr. Wayne now at the Royal Institute of Technology in Stockholm, has now proposed a date that is more palatable to archaeologists. On the basis of DNA from several wolf populations and from the hairs collected off 654 dogs around the world, Dr. Savolainen calculates a date for domestication either 40,000 years ago, if all dogs come from a single wolf, or around 15,000 years ago, the date he prefers, if three animals drawn from the same population were the wolf Eves of the dog lineage." Nicolas Wade: "From wolf to dog, yes, but when?" *New York Times*, November 22, 2002. Per Jensen also suggests this date in a recent scientific book he edited, *The Behavioural Biology of Dogs*. Oxfordshire, UK: Cabi Publishing: 2007.

43 "Humans domesticated dogs": Colin P. Groves, "The advantages and disadvantages of being domesticated." *Perspectives in Human Biology*, vol. 4 (1): 1–12, 1999.

45 He points out: New research pushes the date of the first discovery of a dog back in time. The latest is from MSNBC, reporting on DNA studies at an archaeological find at Goyet cave in Belgium from the Upper Paleolithic period. Jennifer Viegas on October 17, 2008, reporting on the *Discovery News Channel,* writes, "An international team of scientists has just identified what they believe is the world's first known dog, which was a large and toothy canine that lived 31,700 years ago and subsisted on a diet of horse, musk ox and reindeer, according to a new study. The discovery could push back the date for the earliest dog by 17,700 years, since the second oldest known dog, found in Russia, dates to 14,000 years ago." She also reports that ancient 26,000-year-old footprints made by a child and a dog at Chauvet Cave in the Ardèche in southern France support the pet notion. Torch wipes accompanying the prints indicate the child held a torch while navigating the dark corridors accompanied by a dog. Elizabeth Dodd discusses this hypothesis (it is not yet proven) in an online article in *Notre Dame Magazine*, called "The artists of La Grotte Chauvet." See www.nd.edu/~ndmag/w0506/dodd.html. Dodd writes: "But it's also possible he wasn't quite alone. Close to his own tracks are those of a canid—a wolf or a large dog—that may have accompanied him. The researchers can't yet be sure: They haven't found a place where the prints are superimposed, one upon the other, a pattern which would show that they walked together, that afternoon, perhaps it was, or early evening, some 26,000 years ago. It's possible the trails are unrelated, recording visits that occurred days or years or centuries apart. But it's also possible—imagine it—that the tracks are intermingled traces of a joint visit. Studies of the canid prints show they're not quite wolflike; the length of the middle toes seems more in keeping with the proportions we know today in dogs. But the earliest domestication of the dog we've known before now dates to only 14,000 years ago. If all these conjunctions hold up: that the torch marks that have been definitely dated were made by that Paleolithic child; that the two traveled together, beast and boy; and that the beast had, in

the language of Rudyard Kipling, become First Friend, no longer the Wild Dog of the Wet Woods."

45 A recent scientific article: "From wild wolf to domestic dog," by Jennifer A. Leonard, Carles Vilà, and Robert K. Wayne. *The Dog and Its Genome.* Edited by Elaine A. Ostrander, Urs Giger, and Kerstin Lindblad-Toh. Cold Spring Harbor, NY: Cold Spring Harbor Laboratory Press, 2006, pp. 95–113.

45 The first is by David Paxton: The thesis was for the Australian National University. I have not been able to get a copy, but Jonica Newby (also a vet) has read it and talked with Dr. Paxton at length and reports on it in her excellent book *The Pact for Survival: Humans and Their Animal Companions.* Sydney: Australian Broadcasting Corporation, 1997. The book is dedicated to Dr. Paxton.

46 "It was accompanied": Jon Franklin, *The Wolf in the Parlor: The Eternal Connection Between Humans and Dogs.* New York: Henry Holt & Co., 2009, p. 124. Elsewhere in the book (p. 254) he writes, "When the wolf became the dog it lost 20 percent of its brain mass. This we attributed, reasonably enough, to human influence. So why hadn't we immediately attributed the human's loss of brain mass to the propinquity of the dog? Why was it so hard to see nature's impartiality? How could the forces at work on the dog *not* be working, simultaneously, on the human? More realistically, each animal changed the other." He believes that just as humans specialize in thought, dogs specialize in emotions. I agree completely. Dogs definitely are our superior when it comes to feelings.

46 She describes this: Meg Daley Olmert, *Made for Each Other: The Biology of the Human-Animal Bond.* New York: Da Capo Press, 2009, p. 77. She also pointed to a study in 2003 showing that "when eighteen men and women interacted with their dogs [talking to them and gently stroking them] the owners' blood levels of oxytocin almost *doubled*—and their dogs were also twice as enriched with oxytocin!" (p. 74). She recommends the book by Kerstin Uvnas-Moberg, *The Oxytocin Factor.* New York: Da Capo, 2003. It is indeed a remarkable book, making the argument that research has focused far too long on the fight/flight reaction (vasopressin), whereas the hormone oxytocin not only circulates through the bloodstream, it is also found in the brain, as a neurotransmitter. Most important, it is found "entirely unchanged chemically, in all species of mammals." Dr. Uvnas-Moberg is an MD, PhD researcher at the famed Karolinska Institute in Sweden, and considered to be the world's leading authority on

oxytocin. Most of her research was conducted (painlessly, she insists) on rats. She notes that when given oxytocin, even male rats become more gregarious, less aggressive, and engage in friendly socialization. They like to sit next to one another, which leads in its turn to the release of still further oxytocin. I highly recommend the book.

46 Number four is: Wolfgang M. Schleidt and Michael D. Shalter, "Coevolution of humans and canids: an alternative view of dog domestication: Homo homini Lupus?" *Evolution and Cognition*, 57 (2003), vol. 9, no. 1.

47 "Indeed, it can even be argued": Paul Taçon and Colin Pardoe, "Dogs make us human." *Nature Australia*, Autumn 2002, pp. 53–61. It would seem that Aboriginal Australians were the first to use this evocative phrase. I am grateful to Professor Taçon for sending me his article and for the reference to this important phrase.

47 Soon it will be time: Many of the other important recent scientific questions about the human-dog relationship have been addressed in a lively article by Virginia Morell, "Going to the dogs." *Science*, vol. 325, August 28, 2009, pp. 1062–65.

48 That is because: I highly recommend the insights of Randy Malamud in his book *Reading Zoos: Representation of Animals and Captivity* (New York: New York University Press, 1998), as someone who shares my view that zoos are unpalatable.

51 In fact, he would go so far as to say: We had a telephone conversation in Coff's Harbor, Australia, in September 2008. Professor Ruvinsky is the editor of the most important book on dog genetics, *Genetics of the Dog*. Cambridge, MA: Cabi Publishing, 2001.

51 For example, the only domestic animal: Richard B. Lee and Richard Daley, eds. *The Cambridge Encyclopedia of Hunters and Gatherers.* Cambridge: Cambridge University Press, 1999, p. 244.

52 Over thousands of years: Steven Mithen in a fascinating article on "The hunter-gatherer prehistory of human-animal interactions" (*Anthrozoos*, 12, 195–204, 1999), argues that *Homo sapiens* has "gone beyond the predator, competitor and prey relationship that other animals experience into a whole new world of inter-species interactions that are unique to ourselves." I would say: not quite. Dogs also engage in this very same relationship. It is, though, unique to our two species.

52 Wolfgang Schleidt: Wolfgang M. Schleidt and Michael D. Shalter,

"Coevolution of humans and canids: an alternative view of dog domestication: Homo homini Lupus?" *Evolution and Cognition*, 57 (2003), vol. 9, no. 1.

56 It then sits down: See Colin Woodard's report, "Clever Canines," in *The Chronicle of Higher Education,* April 15, 2005.

56 Research by Michael Tomasello: Brian Hare and Michael Tomasello, "Human-like social skills in dogs?" *Trends in the Cognitive Sciences,* 9(9), 439, 2005. See also Brian Hare, Michelle Brown, Christina Williamson, and Michael Tomasello (2002), "The Domestication of Social Cognition in Dogs," *Science*, 298:1634–36, November 22.

58 This might explain: I use quotation marks because there is a movement, which I endorse, to replace this hierarchical term with something kinder, such as "guardian" or "companion."

58 This mutual biorhythm: www.sleepandhealth.com/do-animals-have-sleep-disorders

61 They are literally: I recommend the book *Redemption: The Myth of Pet Overpopulation and the No-Kill Revolution in America,* by Nathan Winograd. Los Angeles: Almaden Books, 2009. See too *One at a Time: A Week in an American Animal Shelter,* by Diane Leigh and Marilee Geyer. Santa Cruz, CA: No Voice Unheard, 2005.

62 "Dogs' understanding": Quoted from "Minds of Their Own" by Virginia Moell. *National Geographic*, March 2008, p. 49.

62 After all: See, for example, David Goode, *A World Without Words: The Social Construction of Children Born Deaf and Blind*. Philadelphia: Temple University Press, 1994, and by the same author, 1992: "Who is Bobby? Ideology and method in the discovery of a Down syndrome person's competence." In Philip Ferguson, Dianne Ferguson, and Steven Taylor, eds., *Interpreting Disability: A Qualitative Reader*, 197–213.

62 We can also stare at dogs: "During much of the time she holds the dog's head tenderly and stares into his eyes." Clifton Sanders and Arnold Arluke, "Speaking for dogs." In *The Animals Reader: The Essential Classic and Contemporary Writings*. Edited by Linda Kalof and Amy Fitzgerald. New York: Berg, 2007.

64 Wolves, we know: An important much-cited article by Adam Miklosi, et al., "A simple reason for a big difference: wolves do not look back at

humans, but dogs do," (*Current Biology*, 13: 763–66, 2003) suggests that the key difference in the personality and behavior of wolves and dogs is that dogs have the ability, and the wish, to look at the human face. This has led to "complex forms of dog-human communication that cannot be achieved in wolves even after extended socialization." No other wild or domesticated animal searches our face with such intensity as do dogs, and this has given them an enormous advantage in understanding our intent, our actions, and most important, our emotions.

65 "But in this life he is dog": *Lost & Found: A Novel.* New York: HarperCollins, 2007.

65 The dog who causes the injury: Marc Bekoff, "Play signals as punctuation: The structure of social play in canids." *Behaviour,* 132 (1995): 419–29. Bekoff showed that some bows carry the message: "I want to play despite what I am going to do or just did—I still want to play." The idea that these are punctuation signals was already suggested by noticing tail wagging in mallard ducks. (J. P. Hailman and J. J. I. Dzelzkalns, "Mallard tail-wagging: punctuation for animal communication?" *American Naturalist*, 108: 236–38, 1974, where the authors believe that the tail-wagging acts as a capital letter or a period, to mark beginning and end of an action.) Bekoff also noted that the vigor of the play bow could actually encourage the partner to want to play. Subsequent research has shown, as was to be expected, and as everyone who has a dog knows, that dogs respond equally to our attempts to play with them: N. J. Rooney, J. W. S. Bradshaw, and I. H. Robinson (2001), "Do dogs respond to play signals given by humans?" *Animal Behaviour,* 61 (4), 715–22. But after playing for many years with dogs, and watching them play with other dogs, I must now acknowledge that any dog worth his salt would rather play with another dog than me, or probably any but the most playful of humans. My experience has been that as they age, dogs seem to prefer playing not with other dogs, but with us. It is interesting that after Konrad Lorenz delivered his first public lecture, his professor asked him if he was always so childish, liking to play. Since Lorenz wrote the first book describing dog play in detail, perhaps he learned about it from his own dogs. Another excellent source for playing in dogs is the new book by Alexandra Horowitz, *Inside of a Dog: What Dogs See, Smell, and Know.* New York: Scribner, 2009.

66 "Perhaps they feel indignant": Marc Bekoff, "Wild justice and fair play: cooperation, forgiveness, and morality in animals." *Biology and Philosophy,* 19: 489–520, 2004. See his new book, with Jessica Pierce, *Wild Jus-*

tice: The Moral Lives of Animals. Chicago: Chicago University Press, 2009. The quote I give is from p. 127. Bekoff wrote to remind me that the dog does not have to actually hurt another dog, but may have just made him uncomfortable or unsure about what's happening, and the play bow lets them know that this was not aggression, predation, or mating, but play. Bowing can immediately precede an action that could otherwise be misinterpreted.

67 Cats, by contrast: John Tierney has a charming article in the *New York Times* (June l, 2009), "In That Tucked Tail, Real Pangs of Regret?" He ends by wondering if cats *ever* regret anything.

67 They lick: Marc Bekoff and J. A. Byers, eds., *Animal Play: Evolutionary, Comparative and Ecological Approaches*. New York: Cambridge University Press, 1998. Also by the same author in 2001, "Social play behaviour, cooperation, fairness, trust and the evolution of morality." *J. Conscious. Stud*. 8(2): 81–90.

68 One way is by barking: See the article by Sophia Yin and Brenda McCowan, "Barking in domestic dogs: context specificity and individual identification." *Animal Behaviour*, 2004, 68, 343–55. The authors point out that dogs can be identified by their bark spectrograms. Barks differ according to context—they are not simply meaningless noise, but an attempt to say something to us and to other dogs.

68 In the last five years: Mirror neurons were discovered by a team of researchers led by Professor Giacomo Rizzolatti at the University of Parma in Italy. See Giacomo Rizzolatti and Lalia Craighero, "The mirror-neuron system." *Annual Review of Neuroscience*, 27 (2004): 169–92.

69 The prominent neuroscientist: See "Mirror neurons and the brain in the vat." January 10, 2006; www.edge.org/3rd_culture/ramachandran06/ramachandran06_index.html.

69 In reporting on the new field: "Cells that read minds." *New York Times*, January 10, 2006.

69 Recently, a video of a dog: See *Huffington Post*, December 3, 2008.

72 Cattle and pigs: Juliet Clutton-Brock, *A Natural History of Domesticated Mammals*. Austin, TX: University of Texas Press, 1987, p. 51. For the notion that people were congregating around permanent sources of water due to climate change, see H. N. Cohen, *The Food Crisis in Prehistory*. New Haven, CT: Yale University Press, 1977.

72 Darwin's cousin Francis Galton: Transactions of the Ethological Society of London, N.S. 3, 122–38. Reprinted in *Inquiries into Human Faculty and its Development.* London: J. M. Dent, 1907. The whole book is available on the Internet. (Alas, it is also the first treatment of "eugenics" and was made use of for nefarious purposes in the next century.)

74 Additional animals: Juliet Clutton-Brock, *A Natural History of Domesticated Animals*, p. 193. Jim Mason, in his excellent book, *An Unnatural Order: Uncovering the Roots of Our Domination of Nature and Each Other* (New York, Simon & Schuster, 1993), shares my concerns.

76 "They usually associate": *The Ecology of Stray Dogs: A Study of Free-Ranging Urban Animals.* Purdue University Press, 2002, p. ix. First published in 1973.

76 The best example of this: Raymond Coppinger is professor of biology at Hampshire College and one of the world's leading authorities on dogs. I am grateful to him for his many emails to me during the writing of this book.

77 "Not a nip, but a full war": Raymond Coppinger and Lorna Coppinger, *Dogs: A Startling New Understanding of Canine Origin, Behavior, and Evolution*. New York, Scribner, 2001, p. 44.

77 The Italian wolf researcher, Luigi Boitani: L. Boitani, et al., "Population biology and ecology of feral dogs in central Italy." *The Domestic Dog: Its Evolution, Behaviour and Interactions with People.* Edited by James Serpell. Cambridge: Cambridge University Press, 1995, pp. 217–56. This inability to go feral may be mutual as well. If I had to survive in the forest on my own, I would be glad to have a dog as a companion. Sadly, humans are all too eager to believe that wolves would be happy to accompany us, as the fraudulent biography called *Misha* demonstrates. She claimed to have survived as a young Jewish girl in the Belgian forests during World War II because wolves befriended and protected her. Only recently did she confess that the whole tale was entirely fabricated, due, she said, to her overwhelming love and affection for wolves (alas, unreciprocated).

80 Of the other animals: I am probably wrong here—many people have no problem with the lives that fish lead in aquariums. This is probably because they believe the myth that fish have only a two-second memory. In fact, fish can remember important matters in their life for over a year, possibly much longer. Life in an aquarium is unnatural and cruel.

82 We don't eat dogs: Some cultures do, of course, and it would seem that eating dogs has been part of every society, at some point or other. Today, however, most Western countries display extreme aversion to the idea of eating dogs, and whenever there is an item in the paper about Korean, Vietnamese, or Chinese dog-eating habits, we wax irate. Animal rights activists are swift to point out that we are hardly in a position to take the moral high road: There are societies where pigs are revered in the same way we cherish dogs, and those societies find our callous disregard for the feelings of pigs just as blameworthy as we find theirs toward dogs. Both are animals with a high degree of sensitivity to pain; both are intelligent; both are affectionate toward us, at least if given half a chance. The point is well taken. Moreover, it is unlikely that *any* society disregards the value of a companion animal per se, and even in those societies where dogs are eaten, the dogs who live with people are not the ones chosen. Many people in all those societies where dogs are eaten raise moral objections to the practice of eating companion animals. See Melanie Joy, *Why We Love Dogs, Eat Pigs, and Wear Cows: An Introduction to Carnism, the Belief System That Enables Us to Eat Some Animals and Not Others.* San Francisco: Conari Press, 2009. See her section on "The problem with eating dogs," pp. 11–16, and on eating dogs in South Korea, pp. 68–70.

83 Remember that even a 50-pound dog: Angela Patmore, *Your Natural Dog: A Guide to Behavior and Health Care.* New York: Carroll & Graf, 1993, p. 23.

85 In the 1950s: In his celebrated book, *So kam der Mensch auf den Hund*, translated into English as *Man Meets Dog.* New York: Kodansha America, 1994, with a new introduction by Donald McCaig; translated by Marjorie Kerr Wilson. F. E. Zeuner, in his 1963 book *The History of Domesticated Animals*, elaborated on the idea of neoteny. I discuss this further in my book *Dogs Never Lie About Love*, pp. 44 ff.

85 This led some to raise the possibility: See, in particular, Stephen J. Gould, "Mickey Mouse Meets Konrad Lorenz." *Natural History*, 88, no. 5 (1979): 30–36 and "Socialization and Management of Wolves in Captivity" by Erich Klinghammer and Patricia Ann Goodmann, in H. Frank, ed., *Man and Wolf: Advances, Issues, and Problems in Captive Wolf Research.* Dordrecht, Netherlands: W. Jung Publishers (Kluwer Academic Publishers Group), 1987, p. 45.

85 There are various names for this idea: See Deborah Goodwin, John Bradhshaw, and Stephen Wickens, "Paedomorphosis affects agonistic

visual signals of domestic dogs." *Animal Behaviour*, 1997: 53, 297–304. The authors show that "physical paedomorphism has been accompanied by behavioural paedomorphism."

87 brains in humans have not evolved: See C. B. Ruff, E. Trinkaus, and T. W. Holliday, "Body mass and encephalisation in Pleistocene Homo." *Nature*, 387 (1997): 173.

87 But what caught the layperson's attention: There are many Web sites where you can view the Axolotl and find out more about this fascinating creature. I would start with Gould, op. cit., p. 177 ff.

87 Thus Temple Grandin: Temple Grandin with Catherine Johnson, *Animals Make Us Human: Creating the Best Life for Animals*. Boston: Houghton Mifflin Harcourt, 2009, p. 36. The title is promising—alas, nowhere in the book does Temple Grandin address the question of how animals make us human. It is clear she was thinking of dogs, but does not say so explicitly.

87 It is true that most wolves: Helmut Hemmer, *Domestication: The Decline of Environmental Appreciation.* New York: Cambridge University Press, 1990, p. 114. "Comparing wild and domestic animals usually reveals decreases in brain size . . . the smaller the brain size of individuals from a species to be domesticated, the more suitable they seem to be for this purpose." Does he mean that it takes a certain amount of stupidity to allow yourself to be domesticated? The evidence concerning brain size is subject to very different interpretations.

89 "What can be absolute is our love": George Steiner, *My Unwritten Books.* New York: New Directions, 2008, p. 176.

90 experts on domestication: *The Decline of Environmental Appreciation*. Cambridge: Cambridge University Press, 1990.

90 Hemmer would explain the floppy ears: Dogs not only hear better than humans, they also hear ultrasonic frequencies that we cannot: "Many animals can hear ultrasonic frequencies; dogs, for example, can hear sounds as high as 50,000 Hz, and bats can detect frequencies as high as 100,000 Hz." Douglas Giancoli, *Physics: Principles with Applications*. Upper Saddle River, NJ: Prentice-Hall, 2002.

91 Dogs, we know, can detect a spoonful of sugar: See *Inside of a Dog: What Dogs See, Smell, and Know*, by Alexandra Horowitz. New York: Scribner, 2009.

94 It is also possible: B. G. Charlton, "The rise of the boy-genius: psychological neoteny, science and modern life." *Medical Hypotheses*, 2006, 67: 679–81. His main point is that "a child-like flexibility of attitudes, behaviours and knowledge is probably adaptive in modern society because people need repeatedly to change jobs, learn new skills, move to new places and make new friends."

95 When I spent some time with [Stephen Jay] Gould: Gould has argued that humans belong to a class of animals in which K selection dominates; that is, we have repeated single births, intense parental care, long life spans, late maturation, and a high degree of socialization. R selected species are animals such as insects who produce thousands of young without caring for them. Gould quotes the philosopher and historian Morris Cohen who wrote that prolonged infancy was "more important, perhaps, than any of the anatomical facts which distinguish *Homo sapiens* from the rest of the animal kingdom." Gould, p. 400. Gould was influenced by a 1962 article by Ashley Montagu called "Time, morphology, and neoteny in the evolution of man," in *Culture and the Evolution of Man* (Oxford University Press, pp. 324–42).

95 Ashley Montagu: *Growing Young.* New York: McGraw-Hill Book Company, 1981. The entire book is about human neoteny. Montagu mentions children's favorite books, such as *The Wind in the Willows, The Little Prince, Charlotte's Web,* and *Ferdinand the Bull,* and says that "it is worth noting that most of these books depend for their success on children's love of animals—a neotonous trait" (p. 141).

96 "The infant's need for love is critical": *Growing Young.* New York: McGraw-Hill, 1981, p. 93.

97 But the mysteries of mother love: I recommend the excellent new book by Sarah Blaffer Hrdy, *Mothers and Others: The Evolutionary Origins of Mutual Understanding.* Cambridge: Harvard University Press, 2009.

97 We are practically alone: Montagu, in *Growing Young*, p. 57, suggests "it was the hunting way of life that almost certainly resulted in the development in humans of the largest number of sweat glands to be found in any animal, a system of glands capable of producing two quarts of sweat per hour over the body surface." Desmond Morris thinks hairlessness evolved because it was more attractive in sexually signaling to both genders. Montagu suggests that dark blue eyes and blond hair were retained because of their paedomorphic value: Babies had them too!

98 Scott and Fuller in a seminal 1965 book: J. P. Scott and J. L. Fuller, *Genetics and the Social Behaviour of the Dog*. Chicago: University of Chicago Press, 1965. For a more modern understanding of the concept of an optimal period for socialization to humans in dog development, see Adam Miklosi, *Dog Behaviour, Evolution, and Cognition*. Oxford: Oxford University Press, 2007, p. 212. The author sums up this way: "If dogs receive no human stimulation before the age of 9–14 weeks, they cannot be socialized. However, there are data showing that even a short exposure to humans can counteract this, and dogs generalize early social experience to other humans. Thus there might be a relatively long sensitive period for developing social relationships with humans."

98 but if we take into account: The figure of two years is disputed. I asked David Mech, the world's foremost authority on wolves, how long they lived, and he wrote to me that "most wolves are dead by five years, but a very few live to thirteen to fourteen in the wild, and seventeen to eighteen in captivity."

99 In fact, this window might extend: Quoted in Gould, op. cit., p. 402.

100 But the very fact: For further reading, see Barry Bogina, "Neoteny: The Evolution of Human Childhood." *BioScience*, vol. 40, no. 1 (January 1990), pp. 16–25. R. C. Fraley and P. R. Shaver (2000), "Adult attachment: Theoretical developments, emerging controversies, and unanswered questions." S. J. Gould, (1979), "Mickey Mouse meets Konrad Lorenz." *Natural History*, 88, 30–36. Based on Lorenz, K. (1943), "Die angeborenen Formen möglicher Erfahrung" [The innate form of possible experience]. *Zeitschrift für Tierpsychologie*, 5, 235–409. James A. Serpell, "Anthropomorphism and Anthropomorphic Selection—Beyond the 'Cute Response.'" *Society & Animals,* 11: 1, Leiden, 2003.

100 Once we were able to: "Recent genetic analyses have suggested that modern dog breeds have a much more recent origin, probably >200 years ago." "Unequal Contribution of Sexes in the Origin of Dog Breeds," by A. K. Sundqvist, et al. *Genetics,* February 2006, 172(2): 1121–28. Darwin was fascinated by dog breeding, especially by the two great groups, herding dogs and guard dogs.

101 *regardless of the effects on the health of the individual dog:* Thomson, K. S. (May-June 1996), "The fall and rise of the English bulldog." *American Scientist*, 220–23.

101 With the odd head: Thomson, op. cit., p. 220.

101 Though entirely subjective: The literature is vast. See among many valuable articles and books, P. D. McGreevy and F. W. Nicholas, "Some practical solutions to welfare problems in dog breeding." *Animal Welfare*, 1999; 8:329–41. Also L. Ackerman, *The Genetic Connection: A Guide to Health Problems in Purebred Dogs.* Colorado: AAHA Press, 1999.

103 Several authors: In a recent book, Michael W. Fox ("America's best-known veterinarian") comes to the same conclusion I do: "Through what I call 'sympathetic resonance' we began to establish social, emotional, and empathetic bonds with other animals, dogs in particular, and we began to regard animals as beings beyond something to kill and eat or wear. . . . So it can be said that humans and dogs . . . began to evolve *together.* Such coevolution includes varying degrees of emotional interdependence, trust, affection, and the ideal of mutually enhancing symbioses where the best interests of both human and animal are realized." Michael W. Fox, *Dog Body, Dog Mind: Exploring Your Dog's Consciousness and Total Well-Being.* Guilford, CT: Lyons Press, 2007, p. 227.

104 new research suggests: Carles Vilà, Peter Savolainen, Jesús E. Maldonado, Isabel R. Amorim, John E. Rice, Rodney L. Honeycutt, Keith A. Crandall, Joakim Lundeberg, Robert K. Wayne, "Multiple and Ancient Origins of the Domestic Dog." *Science,* June 13, 1997, vol. 276, no. 5319, pp. 1687–89. "Mitochondrial DNA control region sequences were analyzed from 162 wolves at 27 localities worldwide and from 140 domestic dogs representing 67 breeds. Sequences from both dogs and wolves showed considerable diversity and supported the hypothesis that wolves were the ancestors of dogs. Most dog sequences belonged to a divergent monophyletic clade sharing no sequences with wolves. The sequence divergence within this clade suggested that dogs originated more than 100,000 years before the present." See the excellent new volume, *Wolves: Behavior, Ecology, and Conservation.* Edited by L. David Mech and Luigi Boitani. Chicago: University of Chicago Press, 2003.

104 It is possible: Humans migrated from Africa about 60,000 years ago. This migration began from a population of between one and ten thousand humans—the ancestors of all living people—and led to both the extinction of Neanderthals in Europe and the disappearance of other human species like *Homo erectus* from other parts of the Old World. Dogs of course went along with the humans in their exodus. In Europe, Neanderthals were replaced by modern humans only between 40,000 and 30,000 years ago.

104 As Robert Wayne points out: Robert K. Wayne, "Molecular evolution of the family dog." *Trends in Genetics*, June 1993 (vol. 9, #6): pp. 218–24.

104 There is no hard scientific evidence: There is an enormous literature on this topic, which I have reviewed in my book *Dogs Never Lie About Love*, but the scientific consensus is that it would be nearly impossible for a wolf to take care of the needs of a small child.

105 Even the most loved wolf: See Raymond Coppinger and Lorna Coppinger, *Dogs: A Startling New Understanding of Canine Origin, Behavior, and Evolution*. New York: Scribner, 2001. "Psychologist and animal behaviorist Erich Klinghammer, director of Wolf Park [in Indiana], is one of the world's experts in taming wolves. . . . but no one at Wolf Park—least of all the handlers—is ever fooled into believing that adult wolves are pets . . . the wolves are only partly tame, and they are still dangerous." This is the reason Janice Koler-Matznick does not believe the gray wolf could have been the ancestor of the dog. See her valuable contribution, "The origin of the dog revisited," in *Anthrozoos,* 15 (2), 2002: 98–118, where she argues that "the most likely ancestor of the domestic dog was a medium-size, generalist canid."

106 This dog, like Immanuel Kant: See Emmanuel Levinas, "The Name of a Dog, or Natural Rights," in *Difficult Freedom: Essays on Judaism.* Translated by Seán Hand. Baltimore, MD: Johns Hopkins University Press, 1990, 1997, p. 153. The passage is remarkable, perhaps unique in Holocaust literature, and deserves to be quoted in full: "There were seventy of us in a forestry commando unit for Jewish prisoners of war in Nazi Germany. An extraordinary coincidence was the fact that the camp bore the number 1492, the year of the expulsion of the Jews from Spain under the Catholic Ferdinand V. The French uniform still protected us from Hitlerian violence. But the other men, called free, who had dealings with us or gave us work or orders or even a smile—and the children and women who passed by and sometimes raised their eyes—stripped us of our human skin. We were subhuman, a gang of apes. . . . we were no longer part of the world. . . . And then, about halfway through our long captivity, for a few short weeks, before the sentinels chased him away, a wandering dog entered our lives. One day he came to meet this rabble as we returned under guard from work. He survived in some wild patch in the region of the camp. But we called him Bobby, an exotic name, as one does with a cherished dog. He would appear at morning assembly and was waiting for us as we returned, jumping up and down and barking in delight. For him, there was no doubt

that we were men. Perhaps the dog that recognized Ulysses beneath his disguise on his return from the Odyssey was a forebear of our own. But no, no! There, they were in Ithaca and the Fatherland. Here, we were nowhere. This dog was the last Kantian in Nazi Germany."

107 Then they would simply gaze at each other: Ted Kerasote, *Merle's Door: Lessons from a Freethinking Dog.* New York: Harcourt, 2007, p. 349.

113 Recent research has shown: Patricia Kaulfuss and Daniel Mills, "Neophilia in domestic dogs and its implication for studies of dog cognition." *Animal Cognition* (2008), 11: 533–56.

115 As Trut concludes: L. N. Trut, "Experimental studies of early canid domestication." In *The Genetics of the Dog.* Edited by A. Ruvinsky and J. Sampson. CABI, 2001, p. 34. The original experiment was first described by the principal investigator, D. K. Belyaev and Dr. Trut: D. K. Belyaev, I. Z. Plyusina and L. N. Trut, "Domestication in the silver fox (*Vulpes vulpes*): changes in physiological boundaries of the sensitive period of primary socialization." *Applied Animal Behaviour Science*, 13: 359–70, 1984/5. Knowledge of the experiment was first widely disseminated by an article by L. N. Trut in *American Scientist*: "Early canid domestication: farm-fox experiment,"87: 160–69, 1999.

117 People make fun: See *One Nation Under Dog: Adventures in the New World of Prozac-Popping Puppies, Dog-Park Politics, and Organic Pet Food,* by Michael Shaffer. New York: Henry Holt, 2009.

119 A tame wolf is not attached: Jozsef Topal, et al., "Attachment to humans: a comparative study on hand-reared wolves and differently socialized dog puppies." *Animal Behaviour*, 7: 1367–75, 2005. The authors, from the department of ethology at Eotvos University in Budapest, suggest that in the course of the domestication of dogs genetic changes took place related to the attachment system of the dog. Wolves simply do not have such an attachment, no matter how socialized.

119 A horse is rarely acquired: See Barbara Jones, "Just crazy about horses: the fact behind the fiction." In *New Perspectives on Our Lives with Companion Animals.* Edited by Aaron Honori Katcher and Alan M. Beck. Philadelphia: University of Pennsylvania Press, 1983, pp. 87–111.

120 they may be grateful: See Paul McGreevy, "Training the opportunist and the comfort-seeker." In *The Finlay Lloyd Book about Animals*. Braidwood, Aus.: Finlay Lloyd Publishers, 2008, pp. 113–25. McGreevy is asso-

ciate professor of animal behavior at the University of Sydney's Faculty of Veterinary Science. He is an expert on horse behavior. My comments about "the release of pressure as the primary reward for horses," come from this article.

122 Dog bites are serious business: The literature is elaborate and complex. Let me just give a few useful references: R. Lockwood and K. Rindy, "Are 'pit-bulls' different? An analysis of the pit-bull terrier controversy." *Anthrozoos*, 1, 2–8, 1987. J. J. Sacks, et al., "Dog bite-related fatalities from 1979 through 1988." *Journal of the American Medical Association*, 262: 1489–92, 1989. L. E. Pinckney and L. A. Kennedy, "Traumatic deaths from dog attacks in the United States." *Pediatrics,* 39: 193–6, 1982. Randall Lockwood, "The ethology and epidemiology of canine aggression." In the *Domestic Dog: Its Evolution, Behaviour, and Interactions with People.* Edited by James Serpell. Cambridge: Cambridge University Press, 1995, pp. 131–38. See too K. L. Overall and M. Love, "Dog bites to humans—demography, epidemiology, injury, and risk." *Journal of the American Veterinary Medical Association*, 218: 1923–34, 2001.

122 We must remember: See the excellent chapter "Gnashing Teeth" by Mark Derr in the second edition of his *Dog's Best Friend: Annals of the Dog-Human Relationship* (Chicago: University of Chicago Press, 2004, pp. 120–57), which includes information on dog fighting. Derr believes that we simply cannot know that some breeds are inherently vicious and untrustworthy. But he does acknowledge that some types of dogs are more inclined to attack a person or another animal without provocation or warning and others are bred to have a very high pain tolerance and aggressiveness toward other dogs. As for pit bulls, he believes they "are best kept out of contact with other dogs, the way some recovering addicts must be isolated from the product to which they were addicted," not very reassuring words for people who believe that pit bulls are no more aggressive than any other breed. It is true, though, as he points out, that no gene for viciousness has ever been found and is unlikely to be. He also makes the interesting point that part of the problem with chow chows, for example, is that they were regularly eaten in their native China, and this could account for the great difficulty just about everyone has in training them.

122 It is difficult to say: See Stephen Collier, "Breed-specific legislation and the pit bull terrier: Are the laws justified?" *Journal of Veterinary Behavior*, 2006, 1: 17–22. He points out "there is no specific research to dem-

onstrate that breeds with a fighting past are more aggressive toward people than other dogs." Even in the five worst breeds, aggressive individuals within these breeds range from 1 in 100 to 1 in 1000, so the individual is far more important than the breed. A recent article suggests that our expectations are often wrong: "More than 20% of Akitas, Jack Russell Terriers and Pit Bull Terriers were reported as displaying serious aggression toward unfamiliar dogs. Golden Retrievers, Labrador Retrievers, Bernese Mountain Dogs, Brittany Spaniels, Greyhounds and Whippets were the least aggressive toward both humans and dogs." "Breed differences in canine aggression," by Deborah Duffy, Yuying Hsu, and James Serpell. *Applied Animal Behaviour Science*, 114 (2008): 441–60.

123 It became clear: See the book by Aphrodite Jones, *Red Zone: The Behind-the-Scenes Story of the San Francisco Dog Mauling.* The real story lies with the humans who "owned" these huge and dangerous dogs. See the article by Mark Derr, "It Takes Training and Genes to Make a Mean Dog Mean." *New York Times,* February 6, 2001.

124 It is hard to talk about aggression: Mark Derr, op. cit., p. 140.

125 But they were wrong: "Saving Michael Vick's Dogs," by Brigid Schulte, July 7, 2008. The cover story of *Sports Illustrated* for December 23, 2008, "What Happened to Michael Vick's Dogs?" by Jim Gorant, is an excellent article. It gives a very vivid picture of what Vick did to his dogs. The article also describes in detail the later lives of many of the dogs. Forty-seven of the fifty-one original dogs were saved! In a recent development, the Humane Society of the United States has decided to use a repentant (?) Vick to speak to minority teenagers who believe fighting dogs is cool and macho. If he can convince them they are wrong, he will have done a service to dogs.

127 It is part of what it means: To understand more about dog fighting, see Rhonda Evans, Deann Kalich, and Craig J. Forsyth, "Dogfighting: Symbolic Expression and Validation of Masculinity." *Sex Roles*, 39: 825–32, 1998.

128 Human fatalities caused by dogs: See J. J. Sacks, et al., "Breeds of dogs involved in fatal human attacks in the United States between 1979 and 1998." *Journal of the American Veterinary Association*, 217, 836–40, 2000. Also: Jeffrey J. Sacks, et al., "Fatal Dog Attacks, 1989–1994," in *Pediatrics*, 97, no. 6, June 1996.

128 And if dogs kill us: The World Federation for the Protection of Ani-

mals estimates that, throughout Europe, some five million dogs are killed each year merely because they are unwanted. I take this from Mary Midgley's wonderful book, *Animals and Why They Matter.* Athens, GA: University of Georgia Press, 1983, p. 103. The current statistics are not perfect. And they can be manipulated. Thus Stephen Budiansky, in his controversial book *The Truth About Dogs*, published in 2000 (p. 185), claims that "some 15 million dogs are relinquished to animal shelters and euthanized each year in the United States; this constitutes about a quarter of the entire owned dog population, and most of these dogs are abandoned because of some intractable behavioral problem, most commonly aggression. These owners certainly did not get what they were looking for in a dog." (He does not provide a source for this figure.) I do not know where Budiansky takes these figures, but according to the Web site of the Humane Society of the United States, "animal shelters care for between 6–8 million dogs and cats every year in the United States, of whom approximately 3–4 million are euthanized." The ASPCA gives comparable figures. If you survey the reasons people give at shelters for abandoning their dogs, most of them are *not* abandoned because of behavior problems, but for reasons of convenience: leaving town, children won't walk the dog, too messy, getting divorced, moving to a new apartment. Aggression is way down the list. The problem is more often with the "owner" than the dog, as any shelter worker will tell you.

128 An even more frightening statistic: For an in-depth discussion, see Kevin Stafford, *The Welfare of Dogs*. Dordrecht, Netherlands: Springer, 2007.

128 There is no end: The Humane Society of the United States has much information on puppy mills. They point out that "Life is particularly bad for 'breeding stock,' dogs who live their entire lives in cages and are continually bred for years, without human companionship and with little hope of ever becoming part of a family. These dogs receive little or no veterinary care and never see a bed, a treat or a toy. After their fertility wanes, breeding animals are commonly killed, abandoned or sold to another mill" (www.humanesociety.org/issues/puppy_mills).

129 Douglas Smith, leader of the Yellowstone Wolf Project: See his *Decade of the Wolf: Returning the Wild to Yellowstone*, with Gary Ferguson. Lyons Press, 2006. At present there are 120 wolves in Yellowstone. They had not been there for seventy-five years. The most authoritative book on the wolf remains David Mech's *The Wolf.* Doubleday, 1970. See too his more recent book, edited with Luigi Boitani, *Wolves: Behavior, Ecology,*

and Conservation. University of Chicago Press, 2003. If I am not mistaken, it was Mech who was the first to point out that wolves did not deserve their reputation for danger to humans.

129 This does not prevent some people: See Jon T. Coleman, *Vicious: Wolves and Men in America*. New Haven, CT: Yale University Press, 2004. "In their stories, Euro-American colonists invented and broadcast a vision of wolves as threats to human safety. They then modeled their behavior on the ferocity they perceived in wolves. Thus folklore explains not only why humans destroyed wolves but why they did so with such cruel enthusiasm" (p. 106).

129 In fact, in the entire twentieth century: I recommend the authoritative account by Mark McNay, *A Case-History of Wolf-Human Encounters in Alaska and Canada* published by the Alaska Department of Fish and Game Wildlife Technical Bulletin 13 (2002) and available as a PDF download on the Internet. In one case he cites, a wolf probably would have killed a small child, but when the wolf released his grip for a moment to get a better hold, a neighbor's golden retriever interposed himself between the wolf and the child. What courage! I bet this docile pet probably never bared his teeth before in his entire life.

129 In Norway: See the essay "Danger from wolves," available at www.wolfsongalaska.org/wolves_humans_danger.html.

129 Bruce Weide notes: www.wildsentry.org/WolfAttack.html.

130 The erasure of knowledge about wolves: See the remarkable book by Brett Walker, *The Lost Wolves of Japan*. University of Washington Press, 2005.

131 "Genuine morality is outraged": Quoted in Mary Midgley, *Animals and Why They Matter*, pp. 51–52.

131 The issue is cruelty per se: This is an entire field of research with a huge bibliography. I recommend the book edited by Frank R. Ascione and Phil Arkow, *Child Abuse, Domestic Violence, and Animal Abuse: Linking the Circles of Compassion for Prevention and Intervention*. West Lafayette, IN: Purdue University Press, 1999.

132 The walk was led by Rob Matthews: Rob Matthews, *Running Blind*. Auckland: HarperCollins, 2009.

133 Not a single one: It has taken science a while to catch up with the popular notion of personality. No one who lives with *any* animal is likely to conclude that personality is the missing ingredient. So it is gratifying to see the following in the scientific literature: "Personality dimensions in nonhuman animals: a cross-species review," Samuel D. Gosling and Oliver P. John, published in *Current Directions in Psychological Science*, vol. 8, 1999, pp. 69–75.

144 Every ten days: The U.S. Advisory Board reported that near-fatal abuse and neglect each year leave "18,000 permanently disabled children, tens of thousands of victims overwhelmed by lifelong psychological trauma, thousands of traumatized siblings and family members, and thousands of near-death survivors who, as adults, continue to bear the physical and psychological scars. Some may turn to crime or domestic violence or become abusers themselves" (U.S. Advisory Board on Child Abuse and Neglect, 1995 report, "A Nation's Shame").

144 An average of nearly four children die: The National Child Abuse and Neglect Data System (NCANDS) reported an estimated 1,530 child fatalities in 2006. This translates to a rate of 2.04 children per 100,000 children in the general population. NCANDS defines "child fatality" as the death of a child caused by an injury resulting from abuse or neglect, or where abuse or neglect was a contributing factor.

145 The fact that in a village in Zanzibar: Budiansky, op. cit., p. 25.

145 There is the excavation: The Natufians were hunter-gatherers, and were the first peoples to live in circular dwellings in what were perhaps the earliest permanently settled villages.

145 There is no way of knowing: See S. J. M. Davis and F. R. Valla, "Evidence for the domestication of the dog 12,000 years ago in the Natufian of Israel." *Nature* 276: 608–10, 1978. A similar early burial of dogs with humans is described in the Hayonim Terrace by E. Tchernov and F. Valla, "Two new dogs and other Natufian dogs from the southern Levant." *Science*, 24 (1997): 65–95. The first agricultural societies followed the Natufian in the Near East, where sheep and goats were first domesticated about 8,000 years ago. This suggests, yet again, that the domestication of dogs could well have given an impetus to further animal domestications.

146 To some extent it is old: See *The Animal Kingdom in Jewish Thought* by Shlomo Pesach Toperoff (Jason Aronson: Northvale New Jersey, 1995, p. 49).

146 Even the Talmud and Midrash: I recommend Boria Sax's fascinating book, *Animals in the Third Reich: Pets, Scapegoats, and the Holocaust*. New York: Continuum, 2000. But he does not comment, specifically, on German shepherds used in concentration camps, and I have not been able to find anything written on this topic. See too Kenneth Stow, *Jewish Dogs: An Image and Its Interpreters*. Stanford Studies in Jewish History and Culture. Palo Alto: Stanford University Press, 2006.

146 an authoritative opinion: "Dog attack deaths and maimings, U.S. & Canada September 1982 to November 13, 2006." Available at many sites on the Web, e.g., en.wikipedia.org/wiki/Dog_attack. According to the Clifton study, pit bulls, Rottweilers, Presa Canarios, and their mixes are responsible for 74% of attacks that were included in the study, 68% of the attacks upon children, 82% of the attacks upon adults, 65% of the deaths, and 68% of the maimings. For more statistics and other studies, see www.dogbitelaw.com/PAGES/statistics.html. Clifton's conclusion is reasonable: "Temperament is not the issue, nor is it even relevant. What is relevant is actuarial risk. If almost any other dog has a bad moment, someone may get bitten, but will not be maimed for life or killed, and the actuarial risk is accordingly reasonable. If a pit bull terrier or a Rottweiler has a bad moment, often someone is maimed or killed—and that has now created off-the-chart actuarial risk, for which the dogs as well as their victims are paying the price."

148 "Some qualities": Jerome Kagan, *Three Seductive Ideas*. Cambridge, MA: Harvard University Press, 1998, p. 76.

149 We can: The scientific article is entitled "Analysing breed and gender differences in behaviour," published in *The Domestic Dog: Its Evolution, Behaviour and Interactions with People*, edited by James Serpell. Cambridge University Press, 1995. The popular book that came out of this research is by B. L. Hart and L. A. Hart, *The Perfect Puppy: How to Choose Your Dog by Its Behavior*. New York: W. H. Freeman, 1988. S. Coren, *Why We Love the Dogs We Do: How to Find the Dog That Matches Your Personality*. New York: Free Press, 1998. A recent Swedish study examined 15,329 dogs and claims to have come up with a shyness/boldness axis, and what they call human "supertraits," a combination of extraversion and neuroticism. I am skeptical. "Personality traits in the domestic dog." *Applied Animal Behaviour Science*, 79 (2002): 133–55.

152 How, if humans and dogs evolved together: See E. A. Lawrence, "Those who dislike pets." *Anthrozoos* 1 (3): 147–48, 1987.

153 Led by the county police chief: www.msnbc.msn.com/id/14139027.

154 So I asked my old friend: On pariah dogs in India, see the excellent Web site: indianpariahdog.blogspot.com.

158 The only thing dogs steal: There is a group in Bali trying to help these dogs by spaying and neutering them, and then returning them to the streets. See www.balistreetdogs.com/about. Note though that these experts say the dogs have never had contact with humans, are entirely wild, and cannot be handled. I believe they can be approached and still retain the potential to be like other dogs if given the chance.

158 One later Muslim tradition: *Encyclopedia of Religion and Nature*, s.v. "Dogs in the Islamic Tradition and Nature." Khaled Abou El Fadl. New York: Continuum International, 2004. Retrieved from the Web at www.scholarofthehouse.org/dinistrandna.html.

Richard C. Foltz, *Animals in Islamic Tradition and Muslim Cultures*. Oxford: One World, 2006. See the chapter "Muslim attitudes towards dogs," pp.129–43. Note p. 130: "The notion exists, supported by a weak hadith, that black dogs in particular are demons in canine form. According to some reports, Muhammad said that a dog or a woman passing in front of a Muslim man praying would nullify his prayers (although the Prophet's favorite wife, Aisha, protested vigorously against this demeaning association)." He also quotes (p. 134) the story of Majnun, the young man who went crazy with love for the beautiful Layla, living alone in the desert, with a great love of dogs.

159 When I was in Australia: Anthropologist M. J. Meggitt, "The association between Australian Aborigines and dingoes." (In *Man, Culture and Animals*, edited by A. Leeds and A. P. Vayda. Washington, DC: American Association for the Advancement of Science, 1965), reports that Aborigines would tame dingo pups who, once sexually mature, would leave the village: "The Aborigines knew this pattern, and women who took a fancy to a particular dingo would break its front legs, so it couldn't return to the wild." But whether this is true is hard to verify. See too A. Hamilton, "Aboriginal man's best friend?" *Mankind*, 7 (1972): 256–71.

159 For the Yarralin: D. B. Rose, *Dingo Makes Us Human: Life and Land in an Aboriginal Australian Culture.* Cambridge: Cambridge University Press, 1992.

160 Recent genetic research: See Emma Young, "Wild dingoes descended from domestic dogs." NewScientist.com news service, September 29, 2003.

160 These visitors: Laurie Corbett, *The Dingo in Australia and Asia*. University of NSW Press (Australia), 1995, rpt. 2001, p. 22. But note: An important recent article by Peter Savolainen and others shows that dingoes are well separated from modern domestic dogs genetically and claim that the minimum separation from dogs is about 6,000 years to a maximum of 12,000 years. "A detailed picture of the origin of the Australian dingo, obtained from the study of mitochondrial DNA," Peter Savolainen, et al., *Publications of the National Academy of Sciences*, August 17, 2004 (vol. 101, no. 33, pp. 12387–90).

161 The dingo is an important member of the family: Carl Lumholtz, *Among Cannibals: An Account of Four Years' Travels in Australia and of Camp Life with the Aborigines of Queensland*. London: Murray, 1889.

161 Laurie Corbett points out: Laurie Corbett, *The Dingo in Australia and Asia*. University of NSW Press (Australia), 1995, rpt. 2001, p. 170.

163 dogs are trained to listen: See D. P. Valentine, M. Kiddoo, and B. LaFleur, "Psychosocial implications of service dog ownership for people who have mobility or hearing impairments." *Social Work in Health*, 19 (1): 109–25, 1993. Also: L. A. Hart, R. L. Zasloff, and A. Benfatto, "The pleasures and problems of hearing dog ownership." *Psychological Reports*, 77: 969–70, 1995.

163 pick up the phone: See S. Duncan, "Service dogs for people with severe ambulatory disabilities." *Journal of the American Medical Association*, 276 (12): 953–54, 1996.

163 turn light switches on and off: See Wendy Aron, "Animals, Canine companions help people deal with disabilities." *New York Times*, November 26, 2006. One woman quoted in the article says, "It would take me fifteen minutes to get out of the bathroom before I had a dog; with Nello, it takes three seconds." On dogs who help people with difficulties in walking in general, see K. Allen and J. Blascovich, "The value of service dogs for people with severe ambulatory disabilities: a randomized controlled trial." *Journal of the American Veterinary Medical Association*, 275, 1001–06 (1996).

163 dogs can scent hypoglycemia: See K. Lim, et al., "Type 1 diabetics and their pets." *Diabetic Medicine*, 9 (2): S3-S4, 1992. Also visit the Web site: www.dogs4diabetics.com.

163 by gently touching the rigid muscles: See the report by CNN in 1997 at: www.cnn.com/HEALTH/9712/29/parkinsons.dogs/index.html. Also

K. Earles, "New hope for Parkinson's patients, service dogs." *Dog & Kennel*, June 1998, p. 36.

163 the dog leads the way back home: For an overview, see C. Wilson and D. Turner, eds., *Companion Animals in Human Health*. Thousand Oaks, CA: Sage Publications, 1998. Israel may be the first country to breed and train these dogs (using only female dogs as the trainers believe they have less ego!) in 2003 (www.jewishmonmouth.org/page.aspx?id=50292): "We know that in this project we are only working with female dogs and not with males. It is important to us that the maternal instinct be present, that they have good eye contact and the desire to please. With the males, their heads are in the clouds or their own egos." Note what the person says about the dog, Polly: "Since I have her, I haven't been afraid to go out to fall or get lost. Because of her I feel free, I'm not dependent on my wife and kids and for months I haven't had to use my cell phone to call on them for help."

163 From the accounts I have read: On the value of service dogs in general, see N. Sachs-Ericsson, et al., "Benefits of assistance dogs: A review." *Rehabilitation Psychology*, 47, 251–77 (2002). Also, for a useful review with all citations, see Deborah L. Wells of the Canine Behaviour Centre at Queen's University in Belfast, "Domestic dogs and human health: An overview." *British Journal of Health Psychology*, 12 (2007), 145–56.

164 The results have been extraordinary: See L. K. Bustad, "Prison programs involving animals." In L. K. Bustad, ed., *Compassion: Our Last Great Hope*. Renton, WA: Delta Society, 1990. Also: L. M. Hines, "Pets in prison: A new partnership." *California Veterinarian*, 5, 7–11 (1983). Also E. O. Strimple, "A history of prison inmate-animal interaction programs." *American Behavioral Scientist*, 47, 70–78 (2003).

164 One has only to look: www.pathwaystohope.org/prison.htm.

164 I am convinced: The discovery of the ability to alert to a seizure was quite by coincidence: S. Pfaumer, "Seizure-alert dogs." *Dog World*, 1992, January, pp. 42–44. See Deborah Dalziel, et al., "Seizure-alert dogs: a review and preliminary study." *Seizure*, 12: 115–20 (2003). The authors describe it as an innate ability to alert and/or respond to seizures. Most dogs are probably merely frightened.

164 seizure-alert dogs: See the article in the *New York Times* of October 31, 2009, by Sarah Kershaw, "Good Dog, Smart Dog," which reports on new research into the ability of dogs. She writes: "Hungarian researchers reported in a study last year that a guide dog for a blind and epileptic person became

anxious before its master suffered a seizure and was taught to bark and lick the owner's face and upper arm when it detected an onset, three to five minutes before the seizure. It is still somewhat mysterious how exactly dogs detect seizures, whether it's by picking up on behavioral changes or smelling something awry, but several small studies have shown that a powerful sense of smell can detect lung and other types of cancer, as the dogs sniff out odors emitted by the disease." See Michael McCulloch, et. al., "Diagnostic accuracy of canine scent detection in early- and late-stage lung and breast cancers." *Integrative Cancer Therapies*, 5 (1): 2006, pp. 30–39. The authors were able to train ordinary household dogs to accurately distinguish breath samples of lung and breast cancer patients from those of controls. For lung cancer, the rate of correct diagnosis (confirmed by a later biopsy) was 99 percent!

166 "My whole life changed": Belinda's experience is typical of people fortunate enough to have these altruistic animals in their life. Stephen W. Brown and Val Strong from the Developmental Disabilities Research and Education Group at the University of Plymouth in the UK write, "An unexpected finding in our early work was that with continued use of a seizure-alert dog, the person's seizure frequency was often reported to show an improvement." "The use of seizure-alert dogs," *Seizure*, 10: 39–41, 2001, p. 40. At the end of the article they report a client with a trained support dog called Rupert: "Before I had Rupert, I had a lot of epilepsy and a little bit of life. With Rupert I now have a lot of LIFE with a little bit of epilepsy." Belinda is now almost entirely free of seizures! See too: "Effect of trained Seizure Alert Dogs on frequency of tonic-clonic seizures." *Seizure*, 11: 402–5, 2002. One of the first articles is by A. T. B. Edney, "Companion animal topics: dogs and human epilepsy." *Veterinary Record*, 132: 337–38, 1993. With children who have epilepsy. the benefits are multiplied: See A. Kirton, et al., "Seizure-alerting and -response behaviors in dogs living with epileptic children." *Neurology*, 62, 2303–2305 (2004). In this article there is a remarkable story of a Great Pyrenee dog who would not leave his three-year-old child for two hours before the seizure. Is it significant that 80 percent of the dogs were female?

168 Sometimes a dog will stare intently: See Val Strong, Stephen Brown, and Robin Walker, "Seizure-alert dogs—fact or fiction?" *Seizure*, 8: 62–65 (1999). The authors, in this article from the British Epilepsy Association, point out that "Support Dogs, a registered charity in England has successfully trained several seizure-alert dogs."

168 But I find it hard to believe: Of course I am not suggesting that altru-

ism in humans and dogs is identical. Earlier I spoke about a golden retriever who put himself into mortal danger by stepping in between "his" toddler and a wolf who had just attacked the child. Humans would probably do something similar. But there are altruistic acts that only humans can perform, for example, harboring a slave in the Underground Railroad. We have the idea that slavery is wrong, and we do not expect dogs to share this idea, but they behave in ways that make us realize they do not accord importance to skin color, which is not entirely dissimilar. There are humans opposed to genocide, who say, "The killing of this particular group is something I want to oppose, even though I may die doing so." This, I agree, is not an idea that would occur to any other animal.

168 *CBS Evening News*: January 2, 2009.

169 You can see this: www.cbsnews.com/stories/2009/01/02/assignment_america/main4696340.shtml. Needless to say, the two-and-a-half minute video went viral on the Web, and can be found everywhere.

170 "If freed they die": Paul Shepard, "On animal friends." In *The Biophilia Hypothesis* edited by Stephen R. Kellert and Edward O Wilson. Washington, DC: Island Press, 1993, p. 287. See too his *The Others: How Animals Made Us Human*. Washington, DC, Island Press, 1996.

170 These are strong words: Recently there has been a great deal of interest in the comparison between slavery and domestication. Aristotle first made the analogy explicit in a much-quoted passage:

"Tame animals are naturally better than wild animals, yet for all tame animals there is an advantage in being under human control, as this secures their survival . . . By analogy, the same must necessarily apply to mankind as a whole. Therefore all men who differ from one another by as much as the soul differs from the body or man from a wild beast (and this is the state of those who work by using their bodies, and for whom that is the best they can do)—these people are slaves by nature, and it is better for them to be subject to this kind of control, as it is better for the other creatures I've mentioned."

Aristotle concluded, in an argument that would have immeasurable influence in Western culture, "it is clear that there are certain people who are free and certain who are slaves by nature, and it is both to their advantage, and just, for them to be slaves." See David Brion Davis, *Inhuman Bondage: The Rise and Fall of Slavery in the New World*. New York: Oxford University Press, 2006, p. 33–34. See too Karl Jacoby, "Slaves by nature? Domestic animals and human slaves." *Slavery & Abolition*, vol. 15 (1994), no. 1, pp. 89–99.

People like William Wilberforce, active in the abolitionist movement in England in the nineteenth century, who was also one of the earliest members of the Society for the Prevention of Cruelty to Animals. See Harriet Ritvo, *The Animal Estate: The English and Other Creatures in the Victorian Age*. Cambridge, MA: Harvard University Press, 1987. See the excellent book *The Dreaded Comparison: Human and Animal Slavery* by Marjorie Spiegel, with an introduction by Alice Walker. New York: Mirror Books, 3rd. ed., 1997.

172 melanoma tumors: H. Williams and A. Pembroke, "Sniffer dogs in the melanoma clinic?" *The Lancet*, 1989, April, p. 734. See too J. Church and H. Williams, "Another sniffer dog for the clinic?" *Lancet*, 358 (2001), p. 930, which provides further corroboration through instances where dogs succeeded in finding tumors missed by ordinary methods. For fascinating and reliable information on the ability of dogs to smell, see Alexandra Horowitz's fine new book, *Inside of a Dog: What Dogs See, Smell, and Know*. New York: Scribner, 2009.

172 The conclusion was positive: Carolyn M. Willis, et al., "Olfactory detection of human bladder cancer by dogs: proof of principle study." *British Medical Journal*, 329: 712, September 25, 2004. Their conclusion: "In summary, our study provides the first piece of experimental evidence to show that dogs can detect cancer by olfactory means more successfully than would be expected by chance alone."

172 The *New York Times* has recently: See Donald G. McNeil, "Dogs excel on smell test to find cancer." *New York Times*, January 17, 2006, reports that Dr. Donald Berry, the chairman of biostatistics at M. D. Anderson Cancer Center in Houston said the claim of a lab in California where the dogs are trained to detect lung cancer and do so with 99 percent perfect accuracy, is "off the charts: there are no laboratory tests as good as this, not Pap tests, not diabetes tests, nothing." More recently, Sara Parker-Pope, in "Dogs sniffing out health problems," *New York Times*, July 6, 2009, reports research from Queen's University in Belfast investigating the ability of dogs to warn their owners of a hypoglycemic attack. There is a British research center, Cancer and Bio-Detection Dogs, with interesting material on their Web site, www.cancerdogs.co.uk. *National Geographic*'s Web site has an interesting video ("Dogs smell diabetic attacks coming") showing the dogs in training.

172 We know, from recent research: "The absence of reward induces inequity aversion in dogs," by F. Range, et al. Proceedings of the National Academy of Sciences, November 3, 2009 (106), 44. Reported by Frans B. M. de Waal.

173 Army dogs saved thousands: See the detailed account of U.S. Army dogs in Vietnam in M. E. Thurston, *The Lost History of the Canine Race.* Kansas City, MO: Andrews & McMeel, 1996.

174 Katz, op. cit., pp. 191–92.

175 This is also the opinion: "Do dogs (*Canis familiaris*) seek help in an emergency?" Krista Macpherson and William Roberts, *Journal of Comparative Psychology*, 2006, vol. 120, no. 2, 113–19.

176 We have all heard remarkable stories: You may recall the phenomenally popular book *The Dog Who Rescues Cats: The True Story of Ginny* (New York: Harpers, 1996), about a small schnauzer/Siberian mix who belonged to Philip Gonzalez, a Vietnam vet living on Long Island, who was himself in a deep depression and despair after an accident until this mongrel came into his life. She was later found to have an extraordinary ability to sniff out and then to save disabled, handicapped, abused, abandoned, sick, or disfigured cats. She even burrowed desperately through broken glass, badly cutting her paws, to find an injured kitten. Why Ginny chooses precisely those animals who would otherwise be shunned is a deep mystery.

178 Do guide dogs know: At least this is the conclusion from the two articles I know that discuss the issue: "The response of guide dogs and pet dogs (*Canis Familiaris*) to cues of human referential communication (pointing and gaze)," by Miriam Ittyerah and Florence Gaunet, published in *Animal Cognition*, 10, 2008. Also by Florence Gaunet, "How do guide dogs of blind owners and pet dogs of sighted owners (*Canis familiaris*) ask their owners for food?" *Animal Cognition*, 11: 475–83, 2008. She concludes that guide dogs do not understand that their owners cannot see them. Marc Bekoff agrees. I asked Peter Beatson, a very distinguished blind professor of sociology from Massey University in New Zealand who has had three guide dogs, what he thought. He wrote me a thoughtful response: "As for whether guide dogs know their clients are blind, I'm sure they don't. At most, they must think we are mentally defective. But all the apparently solicitous care they show for us—avoiding lampposts, stopping at flights of steps, etc., is simply the result of tedious rote learning. They don't, for instance, warn us about stairs because they know we can't see them: They do it because they get yelled at if they forget, and are told they are a 'good dog' if they remember."

178 at the Wolf Science Center in Vienna: www.wolfscience.at/english/wolves/tayanita.

178 "They don't know why": Coppinger and Coppinger, op. cit., p. 258.

179 Their symbiotic relationship: Coppinger and Coppinger, op. cit., p. 27.

179 It is worth remembering: Thorleif Schejelderup-Ebbe, "Beitraege zur Sozialpsychologie des Haushuhns." *Zeitschrift fuer Psychologie*, 88 (1922): 225–52.

180 He may simply be expressing his exuberance: See P. Crowley-Robinson, D. C. Fenwick, and J. K. Blackshaw, "A long-term study of elderly people in nursing homes with visiting and resident dogs." *Applied Animal Behaviour Science*, 47, 137–48, 1996. In one of the elder homes I visited with Benjy, a demented patient was allowed to keep his dog with him at the home. The dog provided pleasure to the man, not to mention many of the other residents as well. I observed the dog often and did not notice any decrease in the dog's quality of life. But nor was I able to see that the dog showed any awareness of where he was and what the people "suffered" from. See too B. W. McCabe, et al., "Resident dog in the Alzheimer's special care unit." *Western Journal of Nursing Research*, 24, 684–96 (2002).

184 "Look, he's grinning!": See Stanley Coren, "Laughing dogs: does your dog enjoy a good joke?" *Modern Dog*, 2 (4), 26–30, Winter 2003.

185 Earlier I quoted: George Steiner, *My Unwritten Books*. New York: New Directions, 2008, p. 176.

INDEX